STORMY

A NOVEL OF CLIMATE CHANGE

Experience the last century as you follow
Stormy around the Earth.

A Novel by
Dr. Richard L. Bailey

Published by TERRA Productions

ISBN 978-0-9890592-1-3

Cover design by Joe Mancino
Jeune Femmeau Hamman by Laurent Hamels – fotolia
Namib by Galyna Andsrushko – fotolia

Change in Annual Averaged Temperature by The National Center for Supercomputing Applications (NCSA), located at the University of Illinois at Urbana-Champaign

Text graphics by the author

Sketches by Myles Dunigan

IN MEMORY OF ELLA AND LEO

SPECIAL THANKS TO AL GORE, JAMES HANSEN, AND TIM FLANNERY

IN LOVING DEDICATION TO MY WIFE AND BEST FRIEND, SUSAN

TABLE OF CONTENTS

PART TWO – THE STRUGGLE BEGINS

CHAPTER FOUR; 2020 - 2030: DIRE WARNINGS

Final Thoughts

GRAPHIC - CLIMATE / CIVILIZATION BALANCE

ABOUT THE AUTHOR

<u>**Main Characters**</u>

- Stormy (Marie Amourette): A world news reporter.

- George: The congressman from Illinois.

- TERRA: The Earth's climate.

- Boothby: A sage and mentor to Stormy.

- Cedric the denier.

- Jack: Stormy's son.

- Susana: Stormy's daughter.

- Lara: Stormy's granddaughter.

- Sarpin: A conservative politician.

- Steve and Anez Amourette: Stormy's parents.

- Dr. Henley: Stormy's college professor.

- Climate refugees.

- The Climate Posse: A group of anonymous anarchists.

- Pierre: Stormy's radical uncle from Paris.

THE CLIMATE CHART

YEAR	6,000 BC (Ice Age)	1800	1960	2000	~ 2040	~ 2065	~ 2095?	Future?	Future?	Future?	NOW
CARBON DIOXIDE LEVEL * (ppm)	160	280	315	374	460	522	600	1,000-2,000	2,000	10,000	980,0000
GLOBAL AVERAGE TEMP °F **	47°	56°	57°	57°	59°	62°	67°	70°	70°+	83°	740°
CONSE-QUENCES	--	--	Carbon levels higher than in last 800,000 years	Drought, storms, flooding refugees heat waves, Ice caps melt	Past climate tipping point	Oil wars, disease, sea level rise, famine hurricanes	Civilization un stable	Ice free planet	Sea level rise = 220 ft.	Life support system in crisis	The planet Venus

* Carbon dioxide (CO$_2$) levels have been rising 2 ppm/year since 1995. ** Future correlations of CO$_2$ to temperatures are estimates.

TECHNICAL NOTES – FANTASY OR FUTURE?

Chapter two is based on real events that occurred between 2000 and 2013. Events in future chapters are likely to occur based on current trends and projections when this book was written. These trends and predictions are described in more detail in the AFTERWORD.

Temperatures are presented in degrees Fahrenheit. Atmospheric carbon levels are presented in the standard designation as parts per million (parts of carbon dioxide per million parts of atmosphere).

Future carbon levels are based upon a rise of two parts per million per year, as has occurred between 1980 and 2010, and on a larger increase which will occur as positive feedback loops inevitably accelerate the rise.

PROLOGUE

Stormy - Your Guide to the Future

"For the sake of your loved ones, and especially your children, you should read this book. If you do, I will tell you how our changing climate affected my family, and my life. The world events I shall report are a warning to all those who live in the twenty first century.

I have come to know that climate change is politically unstoppable and economically irreversible.

I may be wrong, and if I am, you and yours will be safe. But if I am even only partially right, climate will assume its place along with famine, disease, war, and drought as a destroyer of all that people have created since our ancestors came down from trees and first started to walk upright. Much is at risk.

My reporting will be based on events that scientists have said are quite likely in our future. If these events seem impossible to you, I urge you to study the issues, and make your own decisions."

TERRA – The Earth's Climate

"I am the climate: Your climate: The climate for everyone who has ever lived, and whoever will live. I control every aspect of your life; the food you eat, the water you drink, the weather that affects your daily life."

"I am powerful: Stronger than your atomic bombs, your political institutions, and your religious beliefs. I care not about you, your farms or cities, your puny public works or the suffering that I may inflict upon you. I am controlled by the laws of physics, which you may understand, but which you may not change. I am subject to many influences, but the levels of carbon you have placed in the air now overwhelm the other, more natural influences. They have made me more intense, variable, and unpredictable. I am like a drunken weatherman, wandering between hurricanes, drought, floods, heat waves, and icy storms. You may witness my wanderings, but you may not control me. I am the climate. Your ignorance of me and your disregard of my power have inflamed me." "Beware."

CHAPTER ONE – THE SUBWAY FLOODS

Our story begins in the year 2050, a time when climate change has already dramatically altered the Earth, and when the decline of civilization has begun to spiral out of control. This decade, beyond which most authors of the early twenty first century are loathe to predict, marks a time when climate chain reactions are increasing greenhouse gas levels and dominating humanity's efforts to constrain them.

The Death of a Utility

Stormy clutched her umbrella while a gust of wind threatened to steal it away as she approached Christopher Street Station. She had taken a job as a climate reporter in 2024 and was in New York City on an assignment. The tunnel at the subway station was 14 feet below sea level, had flooded back in 2012 and 2025, and she wondered if it would survive the huge storm now camped over the metropolis. Several inches of rain had fallen over the last three days, and several more had been forecast. The jet stream, unpredictable as usual, had locked in place.

Mayor Elmo had agreed to join her. Climbing out of the cab, he appeared, pompous, groomed and boisterous, the archetype of a New York City politician. Stormy, mindful of her profession, kept these thoughts to herself.

"Dinner at Sparky's, drinks on the waterfront and more at my place later," was the bet he offered her that *his* subway would not flood as they walked toward the station. Apparently he did not notice, or was ignoring, the wedding ring on her finger.

"I accept the bet, but for dinner only," she replied. "And if you lose, you agree to publicly state that climate change is killing the subway.

"That will never happen," he smiled as they walked down the stairs.

Descending the stairs they could not help but notice a large sign:

"CAUTION: THIS FACILITY IS SUBJECT TO FLOODING."

It was 3:00 pm and high tide was due soon, rising just as the rain increased in intensity. Seven huge pumps had been positioned at low points in the underground system, and they were already running at full power.

Little did Stormy and Elmo know what was going on in a dimly lit maintenance room where a pump operator spoke into his mouthpiece, raising his voice above the roar of the turbine and calling out to his supervisor. "We're okay. The water is dropping about an inch an hour."

"Good," came a voice from the command center operator on the street above. "It's raining cats and dogs up here. Keep those pumps running at maximum."

"Roger, we're moving thousands of gallons a minute. Wait a minute said the pump operator: The water in the subway is rising, and it's a different color; very muddy."

"Hold on," came the reply. "I'm getting a transmission from the other pumps. Water is rising there, too. What's going on?"

"There's been a cave in," crackled another voice over the radio. "A monstrous sink hole has opened up at Sheridan Square above the subway. The outer wall is cracked and water is gushing in."

"We need another pump," came the desperate transmission.

"Not available." We already have every pump in the city here. Tried to get more from New Jersey last week but they're all in use. They've got floods, too."

"What are we going to do?"

"Unless you can stop the rain, it's going to flood again. Try to get all the trains out this time."

Faster now, water rose around the feet of the workers in the subway, flowing like a river from the growing sinkhole above the subway that was gathering water from hundreds of nearby streets.

"Dammit. We're out of here. Shut her down before the water shorts her out," yelled the supervisor.

As they turned off the pumps, TERRA (as the climate beast had come to be known) increased her fury. Six, then ten, then thirteen inches of rain fell onto the streets of New York. Based on old models, the forecast had been wrong again.

As they boarded the crowded train, Stormy and Elmo were unaware of the sinkhole, or the threat it posed to them. Slowly the cars began moving into the dark tunnel, leaving the safety of the station behind. It was crowded and no seats were available. Flickering lights above the handrail were the first warning, spreading fear through the passengers. In an instant, the train ground to a halt, its electric power robbed by rising waters.

"What the hell! I'll fire those bastards!" exclaimed the Mayor.

"Oh no, not again!" thought Stormy as fear gripped her body.

Within seconds, emergency lights blinked on, temporarily calming a growing panic. But the train stood there, motionless below sea level while water swiftly rose around it as a torrent flooded through the subway tube. In minutes, water began leaking in below the doors and rising to their ankles.

With the pumps gone, TERRA's waters quickly claimed their prize, reaching the station ceiling at the lowest point and flooding subterranean tubes with their entrapped trains.

"Can you swim?" joked Stormy nervously as fetid waters reached their knees.

"We've got to get out of here," said a panicked passenger.

"Everybody, remain calm: I'm the Mayor and I'll get you out," yelled Elmo.

"How?" replied the pregnant woman next to him.

"Open the emergency door!" someone shouted.

Someone pulled the red handle and the doors parted, releasing water behind them, which instantly gushed in around their legs. Screams filled the air as water rose to the passengers' waists. The emergency lights went out, trapping them in darkness below the bustling streets of New York. Above, oblivious to the death struggle playing out below them, shoppers blinked in a driving rain and protected their merchandise from the deluge. Below, in the next car, a flashlight blinked on, illuminating a map of the subway system.

"Find an emergency exit!" Stormy heard above the din of screams for help.

"Go right she shouted." "Get out of here before we all die."

Trapped in a crush of people now fearing their death, Stormy and the Mayor could only go with the flow of humanity, stumbling to a narrow, concrete walkway they could only feel with their feet as water reached their necks.

"How far is it to the exit? screamed Stormy."

"I don't know," replied the Mayor. "Keep going."

"Help me," cried a young mother as she struggled to keep her baby above the menacing water.

Fighting to keep her mouth in the life giving air, Stormy could not help the struggling mother. But Elmo, six inches taller, was able to take the child and hold it above his head, tapping the ceiling above. Then, as if by magic, an exit appeared in the walls, filled with escapees and leading to a dim, glowing light of eternity above.

"Thank God," muttered the mayor is they filed out of the manhole onto a flooded street filled with stalled taxicabs. "Stormy, you win the bet. Climate change is dangerous."

Stormy collapsed onto the pavement, looking up at the white sky above through the rain and wishing that her husband George was there to hold her in his strong arms.

Walking uphill to escape the submerged streets, Stormy eventually found a taxi which took her home.

News cameras were filming the next morning and as the storm and tides receded, bodies of those less fortunate were found floating in the abandoned subway. Several hundred people had given their lives to

TERRA, but the climate did not care. Humanity had bet that climate would not change, and humanity had lost.

Unlike the previous subway floods, there was to be no remedy this time; no bond measure; no repair; no grand plans to dig up thousands of streets to replace an antiquated, broken, and undersized drainage system. In Washington, the conservative government vetoed disaster relief funding for this liberal enclave, citing the need for sea level walls around the capital, drought relief in mid-western states and the oil war in the Arctic.

This time there would be no rescue.

A utility that had once carried thousands of people to work every day is now a permanent river below Manhattan, Stormy's report continued. Reports of radioactive alligators, spirits of lost ancestors and deadly snakes feeding on dog size rats fill local blogs while above the fetid water, traffic has become a nightmare as gridlock grips the city and four, six or eight hours pass before people reach their destinations. Desperate, city officials banned private vehicles downtown and borrowed heavily to buy more buses. Further testimony to how badly the subway loss affected the city came when only twenty percent of the seats were filled for the Yankees / Boston baseball game.

In an editorial, Stormy wrote:

> *People are finally beginning to realize that climate change is real and dangerous, very dangerous. For years, an industrial conspiracy has sowed doubt about global warming, but their motives were profit oriented: Profit for the fossil fuel companies, profit for utilities, and profit for their investors. Their campaign of doubt was easily accepted, for people have a natural tendency to deny something that they fear, but cannot control. In the end, individuals have decided to save*

themselves, but they have not saved the subway; nor would they save New York from future ravages of climate change.

CHAPER TWO – THE CLIMATE IS UNSTABLE
2000 – 2010

Atmospheric Carbon Level – 374-394

Average Earth Temperature -57°
Sea Level - + 0.4 inch
Earth's Population - 6 billion people

The Beginning

Although our story began in 2050, Stormy was born in the year 2000 during an unpredicted snowstorm, Her life spanned the worst of climate change, a period when people first ignored the threat, realized the magnitude of it too late, and finally succumbed to an unstoppable force. Challenged by TERRA, her existence was dominated by efforts to resist the physical laws of nature that would eventually take her children and her husband, determine her destiny, and overwhelm planet Earth.

We return now to a time that many of you have already experienced, and proceed to the year 2100, which many of our grandchildren will live to see, but some, falling victim to the ravages wrought by climate change, will not.

A Child is Born

In the green fields a turning, a baby is born
Her cries crease the wind and mingle with the morn
An assault upon the order, the changing of the guard
Chosen for a challenge that is hopelessly hard
And the only single sound is the sighing of the stars
But to the silence of distance they are sworn
Modified from "The Crucifixion" by Phil Ochs

"Well that one knocked out the power," said Steve as lightning blasted a nearby transmission line. "We had better get to a hospital, *right now.*"

Anez groaned as the labor pains became more intense. It was not supposed to be this way. They were on their way to Atlanta when the storm, largely missed by weather forecasters, hit with a fury. To the south, North Carolina had been hit with more than 20 inches of snow, an all time, single storm snow accumulation record. By the time it hit central Virginia, 100,000 people were without electric power, including Anez, pregnant and in labor with her first child.

Telephones were rendered useless by the storm's fury, so in total darkness, they climbed into their sedan, blinking from windblown snow and shivering in the pre-dawn of January. "Don't worry, we'll make it," said Steve, as he frantically read the map. "There is a hospital in Richmond."

"Why didn't the weather people tell us about this?" she said. We could have stayed at home in Washington."

"The weather is changing," he replied. "Unfortunately television stations don't seem to realize that storms are becoming more intense. This climate change thing is way over their heads."

He hit the four way flashers as they merged onto the empty interstate. Visibility was bad, very bad, and he wished he had not had that third Irish coffee at dinner. A sharp pain pierced Anez and the seat below her was wet. Steve didn't know if he should pull over or rush to the hospital. A blue sign appeared briefly in the headlights "Mercy Hospital–This Exit." He pushed down hard on the accelerator and the little four-cylinder engine responded. There it was, the off ramp, but as he turned the wheel, the car spun violently. Snow flew as the car plowed into a four-foot drift, settling softly in the ditch.

"The baby is coming!" she screamed. A strange quiet enveloped them, broken only by her labored breathing. No other cars were on the road. They were alone. "We can do this," he told her as their eyes locked. And

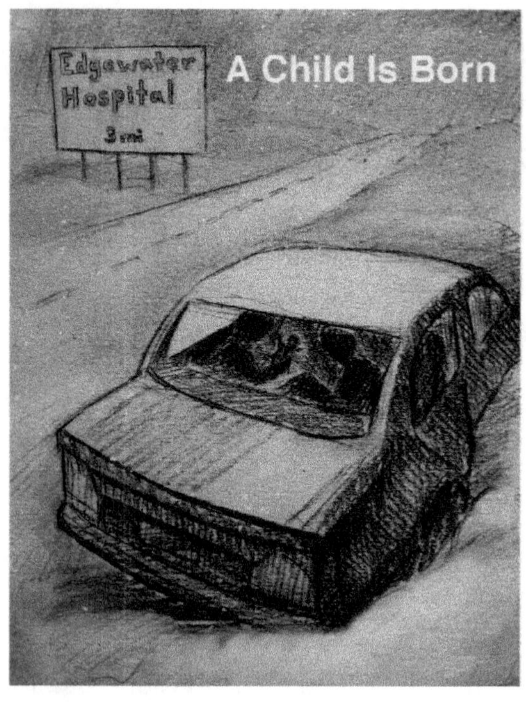

they did. Processes honed by centuries took their course and within minutes the child Marie Amourette was born.

Historians noted that she was one of six billion people on earth, a planet that 50 short years before had only 2.5 billion. An hour later at the hospital, her middle name became "Stormy" destined to live out a century of climate change.

"Perhaps she will be able to do something about these storms," said her father. They almost killed her before she was born."

George

Another baby was born the following year. George Fredericks, the son of a bus driver and a store clerk in Chicago was born at Edgewater Beach Hospital. He swam in Lake Michigan, fished in Canada, and spent summers on a farm in Oklahoma. The fields, woods, and lakes pleased him, but Congress and Washington, D.C became his goal. Perhaps it was the economic depression in 2008 that created this ambition since it caused his father to lose his job and the family to lose their home, something he later vowed to change. But there were bigger forces at work during his lifetime, and although he was intelligent beyond his years, impulsive and witty, George was to be no match for climate change.

And now the News

Two-year-old Stormy crawled into her father's lap as the television announcer described the collapse of an Antarctic ice shelf. A tiny golden retriever puppy barked at the flickering images and shrill voice, as if he comprehended the uncertainty of this event.

2002: ANTARCTIC ICE SHELF COLLAPSES

Scientists were astounded. Such rapid loss of ice had not been expected for decades. Transfixed, the little family watched as satellite images of an area larger than Rhode Island disappeared before their eyes. A scientist who had spent his life studying the arctic was blaming collapse on the five-degree warming that had occurred in the Antarctic region since 1940.

"Don't worry Stormy," said her dad, "it won't hurt us here." Silently, he wondered if that were true.

2003: A HEAT WAVE ENVELOPES EUROPE

It certainly was not true the next year when an unprecedented heat wave enveloped Europe and the ice shelf seemed far away. Stormy was too little to comprehend the event, but was told later that her aunt in France had been one of the more than 30,000 people that had died. Temperature records were broken all across the continent and even at night it was very, very warm. Her father said this was a sign of things to come; that there would be more frequent, and hotter heat waves. It was a scene that she was to cover many times in her coming career as a world news reporter, but as a little girl, Stormy just made her mark on the card that Steve and Anez sent to Uncle Pierre, now trying to cope with the loss of his wife in Paris.

2004: MULTIPLE HURRICANES RAKE FLORIDA

Stormy was too little to remember, but Florida's summer of regret began with Charley, a category 4 hurricane that pummeled central and northeast Florida. Frances, a second category 4 monster quickly followed, spawning 101 tornadoes in her path of destruction. Ivan was next, a bigger, category 5 storm, with a 10–15 foot storm surge and 117 tornadoes.

Mercifully, the season ended with Jeanne, which unleashed only category 3 winds of 111 to 129 miles per hour.

"I'm glad we don't live there" Stormy heard mom remark as scenes of the destruction flashed across the television screen.

2005: KATRINA DEVASTATES NEW ORLEANS
Stormy was five and still remembers the most costly natural disaster in the history of the United States. More than 1800 people died and although she could not understand what it meant, she heard daddy say that damages exceeded 81 billion dollars.

2008: WIDESPREAD CLIMATE DEVASTATION
As little Stormy grew, so also did the destruction from storms fed by ever increasing temperatures and levels of moisture in the air. Hurricanes Ike, Gustav, and Dolly made landfall, destroying crops, oil platforms, and people.

In February, 87 tornadoes were confirmed in southeastern states. In May, 235 tornadoes struck the Midwest/Ohio Valley region. In June, fifteen billion dollars in damages were recorded in Iowa and surrounding states as more than 16 inches of rain fell and 24 people lost their lives.

That summer, severe drought and heat spread across numerous western, central, and southeastern states, destroying crops and bringing lake levels to record lows. As Stormy began school in the fall, thousands of wildfires burned over 5 million acres, destroying hundreds of homes.

2010: HEAT WAVE BAKES RUSSIA: A summer heat wave in western Russia was the most extreme recorded since 1880 for that region. Agricultural losses soared as wheat withered and the nation turned outside its borders for food to compensate for losses from a climate that no one recognized.

2010: PAKISTAN FLOODS
School was more important then, but Stormy heard that one fifth of Pakistan was under water that summer from flooding that affected 20 million people, 2,000 of whom died.

Anez Dreams

White flower petals drifted past her from a large tree above, swirling and coating her shoes. They flickered in the bright, cloudless, blue sky, hiding birds on the branches. Green grass covered her feet like murky, moving water. Then, in an instant, she was flying. The earth below her began to slip away, slowly at first, then faster as fear gripped her heart. She could not get back. Jagged mountains loomed below and a myriad of stars, billions of stars hovered overhead. An ocean appeared, blue, but frothy.

"Wake up honey," said Steve. "You're dreaming again." Mmmmm, they were good dreams," said Anez, "but what do they mean?"

"Well, what did you dream about?"

"There were trees, flowers, and clouds. Then I was flying and there were mountains, the ocean, and stars, billions of stars. "

"Well, why don't we say that mountains are everything people have built, the stars are people, and the oceans ... the oceans are life," said Steve.

"I think the clouds were the sky, and the flowers were the earth, the good earth," replied Anez, as she put her arms around him.

"Unfortunately our good Earth is drifting slowly into climate chaos, and not enough is being done to stop it. But you can't dream any more, I hear a little girl who is hungry."

Stormy Grows Up

That climate would not hurt her also did not seem true several years later when the four hurricanes hit Florida, disrupting the family's annual summer vacation. Even as a small child, the images frightened her. But Stormy was growing into a strong, impulsive, young lady. She splashed in the spring rains, played in sunny summer afternoons, wondered at the colorful leaves of autumn, and kept hearing stories about how she was born in a raging snowstorm. She was home with a baby sitter when her parents went to the movies to see *An Inconvenient Truth* and was outside playing hopscotch when the Nobel Peace Prize was awarded to the United Nations climate scientists. "Maybe things will not be so bad after all," said Anez.

"You're being lulled into sleep," replied Steve. "They're still building coal plants as fast as they can and the world has no policy to stop them. Our little girl will have a very different life than ours."

Predictions

The clock began to strike twelve on New Year's Eve as Stormy, Anez and Steve sat in the living room remembering events of the last ten years. Although some didn't believe it, the world was getting warmer and the amount of carbon dioxide, a greenhouse gas that acts like a heat blanket, was now higher than any time in the last 650,000 years. Methane, another greenhouse gas was rising, too. The earth's average temperature, now at 58 degrees, had already gone up one degree.

"So how can one degree make much difference?" asked Stormy. "Well, if you consider that ten degrees are the difference between our climate now and an ice age, one degree is one tenth of an ice age when twenty-foot high glaciers covered Chicago. Ten degrees warmer would mean than Earth would change as much as an ice age, which is so much that most people could not live here. I know, it's hard to understand," said her dad, "but the one degree warmer that we are seeing now has already changed the weather around the world."

"How?" said the little girl.

"Warmer air can hold more water, which means it can rain more. It also means that the weather can become more extreme, with terrible storms in some areas, drought in other places and heat waves too."

"But it's sunny outside now, and I want to go out and play."

"And so you shall. Don't worry about it," said her father. Stormy thought about it as she swung back and forth on the tire swing. For her, the numbers did not mean anything. But as she grew up, they would come to haunt her.

What she remembered most about what her father said, were the predictions.

"There is a scientist who says if we don't stop burning coal in the next 20 years it will be too late to stop the warming. University professors have said that the world is already changed, and would continue to change even if all the people on earth were to disappear today because the carbon gas stays in the air for about 100 years."

"Why will the Earth change?" asked Stormy.

"Because of all the greenhouse gas that is already in the atmosphere," replied her dad. "We can't take it out of the air. It's like the Earth's temperature is racing to catch up with all the carbon we have already burned." He also told her that many people would not have enough food because the weather patterns would shift, their crops would not grow, and that these people would become climate refugees. These predictions troubled Stormy, but it was time for bed.

"When I grow up I'm going to do something about it. I want to stop all these carbons that are making the climate bad. I'll get a great big fan and blow all those clouds away," she said, holding out her arms.

"That would be a big job; something that only someone like the president of the United States could do. But maybe, just maybe, you can do it."

"All right, I'll become the president," she announced, sleepily.

"Right now though, it is nighty night time," said her dad, giving her three little pats on the back.

Settling in on the couch after Stormy fell asleep, Steve cuddled with Anez, but his mind was troubled. "People don't want to believe that the climate is changing," he said. "Like sheep, they fall for the weatherman's comment that record breaking storms don't mean anything when in reality the climate dice are now loaded; the chances of severe weather are higher and are going to get even bigger. If we wait twenty years to decide that: Oh shit! The planet really is warming: — it will be too late to do anything about it."

"It's the deniers and the corporate disinformation machine," replied Anez. "They know people don't want to believe climate change because then they would have to do something about it. Millions of dollars are at stake. People that would lose that money will say anything create doubt."

"Two American auto companies kept their head in the sand for years, cranking out hummers and SUV's," said Steve. "Look where it got them: bankruptcy."

"It's so simple," said his wife. "The heat gained from carbon in the air is so much greater than any amount we get from changes in the sun. Solar variability effects are overwhlemed by greenhouse gases. The hottest decade ever measured is not from natural changes, it's from us."

"Well, right now you're making *me* hot, so lay back honey. Enough talk about climate."

TERRA Becomes Restless

For eight thousand years TERRA had been stable, comforting and without strife. There had been the usual storms, hurricanes, El Niño's, and floods. Sea level had risen only about six inches in the last century and global temperatures were up only a degree from what they had been in 1980. But these were minor changes, and she had been quiet since the dawn of civilization.

But now there was stress. Carbon levels in the atmosphere were now rising 20,000 times faster than had ever been recorded in the Earth's geological record. Annual increases in carbon that now take one year, took 20,000 years in the past, even when carbon was being emitted by massive volcanoes as the Himalaya mountain chain rose higher and higher.

Like a caged killer being provoked, she kept pacing further, faster, and more frequently. No longer content with centuries of old wind and ocean current patterns, she began to jump the boundaries of these global weather makers. To living things on the earth, the results were disasters not seen or remembered in their collective history. But TERRA, governed by the laws of physics, was unaware, and could not care how her actions played out upon Stormy.

Provoked, almost limitless in potential response and without emotion, TERRA was responding to a continuing and increasing push to add heat and water to the atmosphere. Individual responses played out as events that punctuated normality. More than a thousand people lost their lives when Katrina roared ashore. Millions of trees in America, Canada, and the Soviet Union died and burned as bark burrowing insects survived warmer winters and devastated the forests. Farmers in Australia committed suicide after intense droughts and the homes of Pacific Islanders and Alaskan natives were flooded and lost. But TERRA was just getting started.

The world noticed, but did little. It was a day's newsreel, gone tomorrow. For TERRA, it was just the beginning of an ever-increasing series of what humanity called disasters, but to her were just a response to greenhouse gases that controlled the universe. She would continue her assault.

Stormy Dreams – Fear

Sleeping in her bunk bed, the little girl with pigtails was vaguely aware of how TERRA was changing. Newsreels, adult talk, and the days when storms kept her from going out to play continually reminded her that something was not right. At night, she replayed these things in her dreams.

Fear was a constant response. "The flood is going to get me," she screamed to herself as New Orleans raging waters threatened to overwhelm her. Then she was with the Inuit as their house shook from waves pounding on shores no longer protected by ice. "Will the house fall down?" she asked an old woman dressed in seal skins. "Will I drown?" Cringing in her covers, Stormy tried to hide from TERRA but the climate was everywhere.

Dreaming she was in Africa, Stormy was afraid of starving to death. Swimming with colorful little fish on a coral reef, she watched as they shrank to nothing before her eyes, and suddenly felt herself beginning to shrink. Her tiny body recoiled from the horror, deep beneath the covers.

"Mommy!"

Images of television news reports filled little Stormy's mind. She was in Kansas, looking up at tall skyscrapers that seemed to reach up forever; solid, firm, and indestructible. But the sky darkened, sirens sounded and horrible winds descended from a funnel cloud sweeping toward her in the concrete canyon. A terrible roar obliterated any other sound and the air was filled with swirling objects; trash, trash cans, tree limbs, people, and even cars. Pulled up into the vortex, she looked down as the fifty-story, needle-like building below her slowly twisted on its foundation and collapsed.

Terrified, she awoke in Anez's arms, shaking and wet with perspiration.

"It was just a dream honey," said mom.

"I don't want those bad things to happen," murmured Stormy as she fell back asleep.

But the dream recreated in Stormy's mind that night was to become real in the years to come. A tornado with winds of 350 miles per hour, more powerful than had ever been seen before in human history, would hit Topeka, Kansas, destroying the central business district.

Reassured, she drifted back to sleep, but like TERRA, Stormy's dreams were becoming restless. So many stars glowed in the murky sky that there was little room for darkness. The mountains were still there, but strangely rounded and misshapen. Bright flashes of light flickered in the distance, occasionally coming near. A flower bloomed, and then wilted. A strange dullness covered the ocean. Stormy was afraid.

CHAPTER THREE – CHANGES
2010 - 2020

Atmospheric Carbon Level – 394-415 ppm

Average Earth Temperature -58°
Sea Level - + 1 inch
Earth's Population - 7 billion people

Feedback Loops Set the Stage

Kletka, a native Inuit, was puzzled. Never before in his sixty years of life near Barrow, Alaska had bubbles come up from ponds in the tundra. Yet there they were, hundreds of bubbles that appeared every afternoon when temperatures soared into the '80s. They could be seen in all of the thousands of ponds and lakes that would form every summer.

Having never attended school beyond the sixth grade, he did not know the meaning of this event, now being replicated across the northern latitudes of planet Earth. He did not know that methane, now being released from millions of acres of thawing tundra would inflame climate change more than human civilization had ever experienced.

Halfway across the planet, other greenhouse gases were also being released into the atmosphere as tropical jungles in the Amazon and rain forests in Indonesia succumbed to the need for farmland to feed hungry people. Carbon, released from the jungles of the Earth, joined methane in a frenzied dance of climate change. Kletka did not know that the bubbles he was seeing were making the Earth warmer, which was melting even more permafrost and creating even more bubbles.

Other feedback loops were also in place, gaining strength as they grew and amplified each other. In the Arctic, darker ocean water absorbed more heat, and melted even more ice. Across the plains a scorching sun burned off more carbon from soils and dying vegetation. Relentlessly, like compound interest, the volume of greenhouse gas in the atmosphere grew, then grew again in a spiraling, upward cycle.

20

A thousand miles south, signs appeared along the freeways in Berkeley, California. "Wake up and smell the permafrost" they implored. But hardly anyone did.

Ten Candles

Happy birthday to you. Happy birthday to you. Stormy was ten and bubbling with energy. "You are going to be a writer" was scrawled across the birthday card from uncle Pierre in France after he read her first essay. "Maybe someday she will write about how bad it was for President Obama to open up the arctic ocean to oil drilling," he had confided to Steve and Anez. "You might as well be drilling for greenhouse gases." But Uncle Pierre was always a skeptic, and Europeans took a much more dire view of the energy crisis than Americans. His card was lost in the wrapping paper and forgotten.

At school, Stormy played climate games on the computer. Her favorites were the *Chameleons Go Green* game from New Zealand and games at the NASA website. At school she learned that the carbon she was breathing today would probably be half way around the world next week. "This stuff really gets around," she said. It was hard for her to understand how something so small that you could not see could be changing the whole planet. But then her teacher explained to her that, even though there was only a tiny amount of carbon in the air, there was a whole lot of air, and the sunlight shown through all of it. "If you count all the carbon in the air on earth, it is a really big number!" she told her mom.

Like many children her age, Stormy was learning to turn off lights and walk to school. She remembered hearing about how most of the coral reefs were disappearing around the world. The link between those light bulbs and reef was vague though, and her teacher was not sure it was really as bad as the green groups said it was. At the aquarium exhibit there were a few words about reefs dying, but the coral and fish there were still pretty. A few people shook their heads when they read that the oceans were becoming more acid, but they were much more concerned about their house payments and jobs. Coral reefs would bounce back; they always had.

Summer Vacation

It was hot in San Francisco, 95 degrees hot, when Anez took her there for a summer vacation. Dad was presenting a paper to a science conference and they went to the new museum in Golden Gate Park. At the climate exhibit she wandered between the displays, some declaring that thousands of species were going to die if business as usual continued.

"But what is a species anyway?" she overheard someone say. "Can you eat it?"

Wandering through the exhibition, Stormy wondered why so many exhibits were about light bulbs, the kinds of food you should be eating, and how much better it was for you to have solar panels. What about the people already dying in Africa because the weather patterns have shifted? What about the millions of dead trees in Colorado, Montana, and Alberta? What about the thousands of people who have already died in heat waves and the two billion people in India and China that won't have enough water when the glaciers are gone she wondered.

Why don't they tell people that carbon levels are higher now than they have been since 650,000 years ago, that they are still rising faster than ever before, and if they keep going up it could destroy all life on earth? But the exhibits were like milk toast, and they ignored the real threats.

Another demonstration had a shiny wind generator and people wrote comments about how great it was that technology was going to reduce greenhouse gas emissions. But at age sixteen, Stormy was beginning to have her own ideas that technology might not be able to save us. When writing her high school science paper she learned that carbon dioxide levels in the air were now at a level of 405, but had only been 368 when she was born. Still, nobody had told her what the difference meant, and because few people listened to teenagers, she kept her thoughts to herself as they left the museum and went to a local restaurant.

People were waiting in line for dinner at the Ocean Cafe that night and the busboy was passing out free glasses of wine.

"Red or white?" he asked, carrying a decanter of each. Orange lights glowed overhead, and conversations, enlivened by the wine, filled the room.

Steve was fondly remembering 2010 when wild salmon and rock cod were on the menu. Now however, cod numbers had crashed and wild salmon were disappearing. Populations of many ocean fish had crashed and what was once commonly sold in stores was becoming food for only the very wealthy. Scientists noted that oyster larvae were unable to survive higher acid levels in the oceans. It seemed that global warming was about more than the earth getting hotter. The greenhouse gases were also making the ocean more acid and effecting what Stormy could have for dinner.

Reading the news on dad's tablet between bites of calamari, Stormy said, "Look mom, it says here that the global average temperature is only 58 degrees. It sure feels hotter than that now."

"You need to learn what average is," said dad. "An average includes all the temperatures on the planet, including the north and south poles, where it is really cold. The important thing to remember is that the averages have been getting warmer every year. The decade when you were born was the hottest ten years ever recorded."

"OK, a colder winter in one area doesn't mean the average everywhere for the year is not warmer, and some places can become cooler for a while even though the Earth is warming up. I understand that. But was the Earth's average temperature always 58 degrees?" asked Stormy.

"Nope, when you were born is was one degree colder, 57 degrees."

"Could one degree make the salmon go away?"

"Perhaps. Pass the bread. You have to remember that one degree affects everything, not just the salmon, but also what they eat and where they live."

"Maybe it is because the salmon can't regulate their body temperatures like we can," said mom.

"I can believe that," said Stormy, remembering how it was when she had a fever. "Maybe the earth is getting a fever. If it is so important, why aren't people doing something about it?" she asked her parents.

"I guess not enough people care," said Steve.

"No, it's those timid politicians," said Anez. "All they do is compromise to get votes."

"Yeah, but greenhouse gases don't compromise," said Steve. "Pass the farm fish."

Overhearing their conversation, a grumpy man at the next table commented to his son.

"Don't listen to that crap junior. It's just a bunch of lies that Al Gore made up to make himself rich. The climate isn't warming. They even had record snowfalls in New York last winter."

Stormy watched intently as her father turned to answer the challenge.

"You're one of those deniers," he said. "Those people who let politics talk for them."

"You bet your ass buddy. If you left wing liberals are for it, I'm against it. It's just another excuse for taxes and I didn't elect my tea party person to compromise on global warming."

"Tell that to the 320 tornadoes that hit in 2011," said Steve. Maybe one of them will get you someday."

"Ah that's somewhere else," he muttered.

"Yeah and you don't care about those people somewhere else, do you?" yelled Stormy.

The response from someone half his size disarmed the denier and he turned back to his wine.

"Steve, we have to go," said Anez. "The lecture starts in half an hour."

"God damn scientists," scoffed the drunk as they walked out the door.

"God damn deniers," said Steve. "They'll drive us all to hell."

And Now the News

Glued to the computer screen, Stormy was fascinated by the world around her, even if the news was grim.

"I want to be an international reporter," she exclaimed. "I want to tell people about it and stop this climate change."

"Well then you better pay attention to the news," said mom. The announcer droned on, summarizing the climate events of the decade:

2011: HUGE FLOODS RAVAGE CHINA, INDIA, BANGLADESH, AND PAKISTAN: Flooding continued as glaciers melted and torrential rains sent the Yellow, Yangtze, and Ganges rivers out of their banks. Hundreds of millions of people survived in squalid conditions, awaiting their fate in a Climate-changed world. Many became refugees, seeking solace in whatever country they could find it.

2011: THE MISSISSIPI BREAKS ITS BANKS: In America the Mississippi river reached record levels, broke through old levees and flooded homes, farms, and businesses. The murky water reached levels as high as ever seen in anyone's lifetime.

2012: GLOBAL WARMING DEAD: Senator In-huff (the conservative politician) declared that global warming ended and never really began anyway as reported in the Washington Post. These droughts have happened before, and this one will not last he said. I propose more money from Washington to drill deeper for ground water to plant more corn. We know that corn makes good bio-fuel and bio-fuel means money in your pockets. The crowd cheered.

2012 DERECHO RAKES AMERICA: Driven by straight line winds of 80 to 100 miles per hour, a freak storm called a derecho began in Indiana and roared into Baltimore last night reported WGMedia. Millions of people sweltered in 90-degree heat due to downed power lines which cut off power for days. Across America, more than 3,000 temperature records were broken or tied in June of 2012.

2012: HEAT BAKES AMERICA: "Hottest summer since the dust bowl!" exclaimed the Topeka News. Cracked home foundations from drying soil were only one of the serious consequences of the heat and drought that held mid-western states in its grip. Bankruptcy loomed for small farmers as crops withered and millions of dead fish littered algae laden lakes and streams.

2012: SANDY DEVASTES NEW JERSEY: It was called the storm of a lifetime; a hurricane colliding with a west moving low pressure system and an arctic air mass. A hundred died and millions were without power as the New York subway flooded and Jersey boardwalks disappeared. Even the big apple mayor realized that climate change had made it worse, much worse.

2015: GLACIER MELT THREATENS AGRICULTURE: Glaciers on the "roof of the world" continued to melt, and they will eventually be completely gone predicted Chinese scientists. Then, we will be faced with unprecedented drought and no water for irrigation, like the drought they are now experiencing in Australia he continued. "These glaciers irrigate crops used by one fourth of the world's people: Where will we get our food? It is a question that will be increasingly asked by billions of our people," he said.

2016: CLIMATE CHANGE OPENS THE NORTH-WEST PASSAGE – UNITED NATIONS FAILS TO RESOLVE ARCTIC DISPUTE: Rhetoric increased today when the United Nations failed to resolve competing claims over Arctic oil and natural gas rights reported Reuters. Because the United States had failed to ratify the Law of the Sea Treaty, their claims to land north of Alaska were still in limbo. Meanwhile, exploration efforts by major oil companies intensified as summer ice melted and shipping routes between Asia, Europe and North America opened up for the first time in human history. Norway celebrated the opening of their new natural gas terminal in Siberia, while the U.S. State Department brushed off the legality of Russia's under-sea flag planted below remaining ice by a submarine.

The great Arctic oil rush had begun. It was a headline that was to become all too common in Stormy's life, and one that would eventually take away her son.

> *2017: TAR SANDS OIL FLOWS TO THE GULF: This puts us on the path to uncontrollable climate change, claimed environmentalists as carbon laden oil from Canada began flowing to American refineries in Texas. Blocked for years by law suits and civil disobedience, the controversial pipeline will bolster the profits of oil companies that will sell fuel to burgeoning markets in China and Brazil. The pipeline guarantees that huge deposits of dirty crude oil would be mined and burned for many decades to come.*

Uncle Pierre Arrives

"Viva la France!" he shouted as he disembarked from the airliner, waving a tiny French flag.

Wearing red, inebriated and frisky, Uncle Pierre greeted Steve, Anez, and Stormy. "I don't like this country very much, and I intend to do something about it," he said as they picked up his luggage. "Your health insurance is ridiculously expensive and you're the world's second worst carbon emitter, but look out, Pierre is here."

Taken aback, Steve didn't know what to say, but managed to whisper, "Are all your uncles this crazy?"

"Oh don't worry about him," replied Anez. "He's harmless... at least, I think so."

On to College

As she gave the commencement speech to her high school classmates, Stormy reminded them that, no matter what we do now, carbon already in the air will cause another degree of warming because there is no way we can remove it.

"Look at how much carbon is in the earth," she said, showing a graph on the screen. "Just a small fraction of what is in fossil fuels is enough to plunge the earth into runaway climate change, like Venus."

"And when ice sheets melt and sea levels rise, how will we stop it?" she asked.

"Just re-freeze the arctic!" shouted a heckler. It was Cedric; the jerk that had taken her to the prom but hardly spoke to her. Unlike Stormy, he was a bit of a loner who spent a lot of time playing video games. From his comments during the school debate on global warming, she knew he was a "denier" one who didn't believe in climate change or maybe couldn't understand it. His father worked for an oil company, and was the source of most of his information.

"I'm not afraid of climate change," he said. "We have the technology to create whatever we want. Civilization has already changed the earth more than climate has and we're still doing fine, just fine. People can handle anything this planet can throw at us."

"Oh yeah!" Stormy yelled back. "Is the technology helping the tens of thousands that die every year from malaria, dengue fever, cholera, and other diseases that are spreading because of climate change? And what about famines and wars caused by drought? Can you stop sea level rise for two centuries? And who is going to pay for all of it?"

Stormy was just warming up.

"And just how are you going to re-freeze the whole arctic? And how can we afford to build sea walls around all the shorelines of the earth, like in London, Washington D.C., Miami, Manila, New York, and Hong Kong?"

Realizing that she had been yelling, Stormy moved back behind the podium, staring intently at her adversary.

"Well?" she asked.

Embarrassed, Cedric fell silent, but his buddies picked up the challenge. "It's just a natural cycle, so how are you going to stop it?" yelled out another denier.

"Natural my butt," replied Stormy. Do you call changing the composition of the air natural? We can stop it by switching to solar power, wind turbines, electric cars, and better insulation in houses. And I'm the one who is going to get people to do it, you jerk. Shut up; I'm giving the commencement speech, not you."

Stormy was not about to be intimidated by deniers.

She then asked her classmates, "Why do we have to wait until climate destabilization affects us personally? What about the people who are already suffering, like the Inuit of Canada, or the citizens of south sea islands whose homes are being flooded, or the people starving in Africa?"

She urged them to dedicate their lives to preventing such events, which had become increasingly common during her short life. More than 100 "storms of record" the kinds that produce rainfall or floods thought to happen only once in 200 or 500 years had occurred in the last ten years. As a budding journalist, she wanted to tell people that things were changing.

The ceremony ended, graduates threw their mortar board hats into the air, and everyone filed out of the auditorium into a warm summer evening, where Stormy's argument continued.

"Show me the numbers," shouted Clarence, a freshman attending the graduation with his sister. I dispute those numbers: They haven't been proven."

Stormy turned to face her accuser who was standing there with arms on his hips in a taunting manner. "Oh, so how did you get to be an expert on climate change?" she inquired.

"I know what I know," he replied. "These so called experts are making it all up. They have stupid titles like Head of the Goddard Space Institute and Digital Communications Specialist. There is no way they could know all the stuff they say they do."

"Wow. You know more than people who have spent their whole lives studying it; more than the academies of science around the world and more than the Toyota executives whose factories were just flooded in Thailand. That is either amazing, or just plain dumb. What's your source of information, Fox news, or The Daily Planet?"

"Stormy, calm down, he's just spouting off," said Anez. "Some people have a closed mind."

"When they stop progress, closed minds are dangerous," replied her daughter. "But you're right mom, I've got better things to do than try to educate someone like that. The world is full of peer reviewed articles by learned, honored people. The information is out there. You can lead a horse to water but you can't make it drink. What I want to know is how they think all these scientists reach their conclusions. Do they think they just make it up? Do they see a conspiracy in everything they disagree with?"

A storm interrupted their conversation, pelting the hapless graduates with hail as they ran to their cars. "Our car is using less energy than your car," shouted Stormy as she drove off in her 2002 Prius.

"I don't believe it," was the reply, which few heard as a bolt of lightning flashed overhead.

At home, the war of words continued on the internet.

"Electric car batteries cost so much that it is cheaper to drive a gasoline powered car," tweeted Clarence.

"How convenient of you to ignore the public cost of pollution and climate change that are costing billions of dollars and the lives of millions of people," Stormy responded. "Climate change has costs, for everybody. You're forcing these costs on us just because you want the cheapest deal for you. That's called being selfish."

"Those batteries don't even last long, they wear out," replied Clarence.

"They're guaranteed eight to ten years; what other part of a car is guaranteed that long?"

"Yeah but recharging stations are few and far between," retorted her adversary.

"Not where I live. There are dozens of stations within twenty miles, and more are going in every year."

Becoming desperate to put down electric cars, Clarence continued the argument.

"If you use electricity from dirty coal it can produce more greenhouse gas emissions than gasoline," he wrote.

"You must be talking about China. But I don't live there and my electricity is from 50% renewable sources."

"These electric cars, they'll never catch on. They're a toy for rich liberals."

"Is that why the Prius was the bestselling car in California in 2012?" countered Stormy.

It was midnight, and Clarence didn't have an answer.

Political Strife

But as Stormy was graduating, more than climate was changing. Rallying conservatives in the mid-west, Sarpin, the perennial conservative candidate, was raising funds for Senator Inhuff who was expected to win handily in the next elections. Encouraged by the continuing economic depression and thousands of bankrupt mid-western farmers, the political right had gained strength, and was about to take over local, state, and federal governments.

"No longer will there be money for energy gimmicks or green stuff," said Sarpin, echoing the Reagan administration that had removed solar panels from the white house in 1986. "This climate change is just a natural cycle. Next year it will be cooler."

"Crap!" exclaimed Stormy when she read Sarpin's news column, shaking with anger.

Stormy was not easily angered, but when it happened, she could not get it out of her mind. She became even more upset when she learned that many people believed that climate change was part of a natural cycle, and therefore nothing to worry about. "Some dopes even believe that global warming is not real," she told her dad.

"Some of them don't believe it for political reasons," he responded. "Some because they don't want it to happen and some because they've been told what to believe by television ads which are sponsored by oil and coal companies. But all of them are ignorant because the evidence is overwhelming."

"Can't people understand that the North Pole is melting? Someone needs to tell them." Stormy was still mad.

"Well, that would be something you could devote your life to." Like a tune that you can't get out of your mind, the thought lingered with her, played around the edges of her consciousness and popped out when least expected.

"I'm smarter than these dumb deniers," she thought. "If I can just get enough information out there, they will see how serious this will become. The whole planet is being screwed up and they're out there saying it isn't real. Who is the real enemy here?"

The End of a Decade

As Stormy turned twenty, the United States and indeed the world economy had turned sour. Greed, corruption, and an aging population had created financial deficits throughout all levels of world government. On top of that, damages from oil spills, hurricanes, tornadoes, and floods were beginning to take their toll. It was as if the combination of politics, pollution, and job losses were acting together to limit her potential, and her life.

But for now, college beckoned. Stormy was ecstatic when she was accepted by the University Climate Institute. Saving the world would have to wait.

Business as usual had prevailed in the second decade of the twenty-first century. As the world's population added seven hundred million people and grew to 7.5 billion, climate change was important, but so were many other things. So hardly anyone noticed when levels of methane, a much more powerful greenhouse gas, edged up five percent in the atmosphere, or when global temperatures approached 59 degrees. For civilization, it was business as usual; but for TERRA, it was not.

HOW MANY GREENHOUSE GASES ARE WAITING TO BE RELEASED?
In Billion Tons (Gt) of Carbon Dioxide or Methane. Bar sizes not to scale.

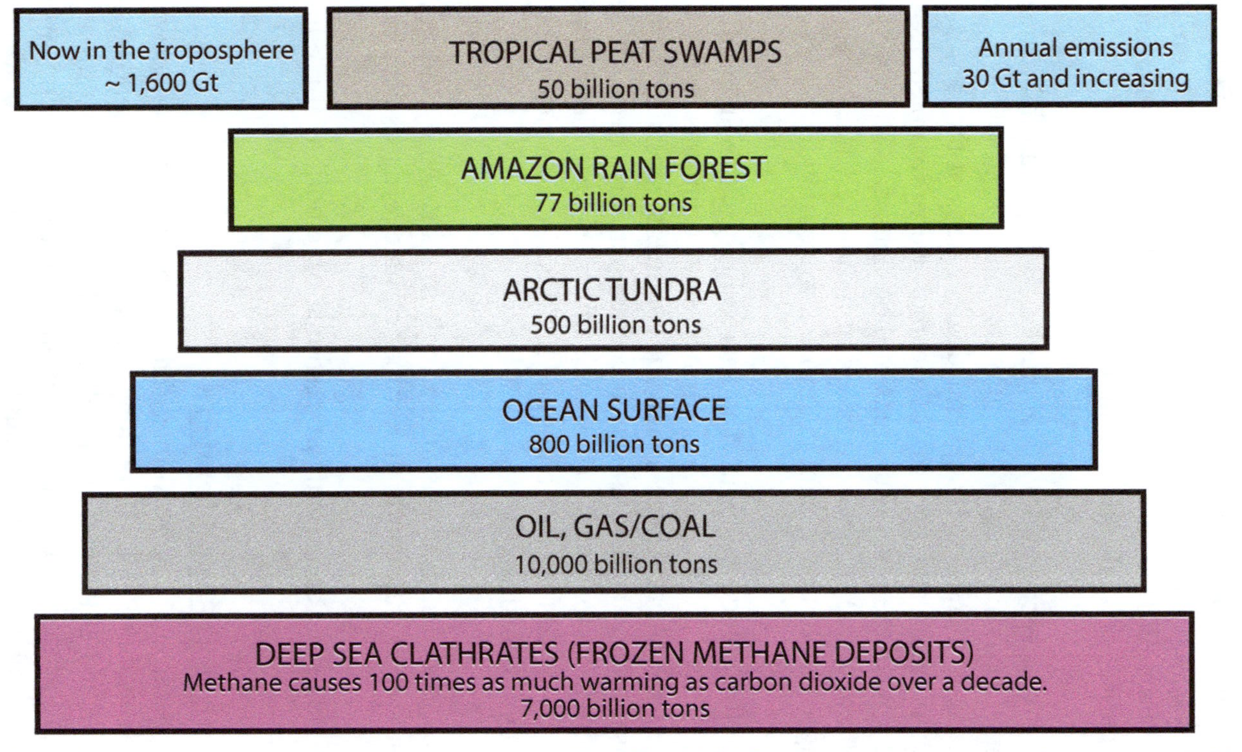

How much could greenhouse gases increase world temperature? The 2,800 Gt of carbon in coal, oil and gas reserves of fossil fuel companies is five times the amount that would increase our earth's average, annual temperature by the agreed upon safe limit of 3.5° F. If all burned, potential temperature increases of up to 15° or more would drastically alter the earth as we know it.

TERRA - Breaking Out

Unrelenting, the forces of greenhouse gases inflamed TERRA. Not only were they increasing dramatically, they were feeding upon themselves, gaining strength from the changes they created. Unchained and unlimited, the amplification of TERRA continued.

At first it was apparent only in averages. Average rainfall increased by 12 percent (thousands died in floods and mud slides); night time temperatures increased by several degrees (tens of thousands died in heat waves); rhythms and patterns of weather were disrupted as jet streams shifted toward the poles (hundreds of thousands died as deserts spread into agricultural areas); glacier fed rivers in India and China dried up (millions of refugees sought new homes in more climate friendly lands).

As TERRA gained strength, she increasingly came in conflict with the works of man. In Russia, half of the wheat crop was lost as she threw unprecedented heat waves across the continent. Tens of thousands of cattle in Kansas and Texas died as she forced temperatures over 100 degrees, and kept them there for weeks. Thousands of her subjects died in Pakistan from floods that lasted for two months and hundreds of factories in Bangkok were closed by floods that year.

Across continents, food was increasingly in short supply and as the African drought spread into southern Europe, climate refugees became a menace.

All of this was done without restraint or emotion, for climate has no conscience.

Stormy Dreams - Apprehension

As she became a young woman, Stormy became apprehensive. She cut her long, auburn hair, took long walks by the sea, played in the surf, and wondered at the masses of jellyfish that would wash up at high tide. Her apprehension was strongest at night when the wind would blow curtains through her open window and tree branches would sway in dark shadows. Sensing TERRA's growing power, her dreams were filled with fear, frustration, and dread.

Walking in fields of dead grain, the darkening sky seemed to close in on her. "Go away…" she murmured, "but there was no response. "You are not real!" she shouted but only images of dying cattle came to her in reply. Half awake, she felt a need for water but it came in muddy torrents with floating houses and bloated bodies.

"What if we can't stop this?" she asked. "What will happen to all of us if we can't lower carbon levels? Where will we go? Who will live and who will die?" Stormy dreamed, but no answers came to her.

CHAPTER FOUR – DIRE WARNINGS 2020 – 2030

Atmospheric Carbon Level – 415-435 ppm

Average Earth Temperature -58.5°
Sea Level - + 6 inches
Earth's Population - 7.5 billion people

Images of innocence charge her to go on
But the decadence of destiny is looking for a pawn
To a nightmare of knowledge she opens up the gate
A blinding revelation is laid upon her plate
That beneath the greatest love is a hurricane of hate
And God help the critic of the dawn.
Modified from "The Crucifixion" by Phil Ochs

Feedback Loops Kick In

"It's too hot."

"So, turn up the air conditioning," was repeated millions of times over and over in the wealthier nations of the world. But as power demands for cool air increased, so did carbon emissions, growing to record levels each year, and forcing temperatures even higher as another feedback loop kicked in.

Taking notes at the world climate conference in London, Stormy listened intently as scientists jointly discussed their latest computer studies.

"Feedback loops have begun and they are making global warming much worse," said the professor at the podium. "Climate change is like sound building up in a microphone; the louder it gets, the more that sound makes it louder."

To demonstrate, a sound technician boosted the amplifier until it shrieked and people covered their ears.

"Feedback is an event where something causes more of itself," continued the speaker. "But, as you know, a positive feedback loop is not necessarily good, it just means that it is increasing, not decreasing, as it would in a negative loop. Feedback loops are self-re-enforcing. They keep amplifying themselves."

The overhead screen translated a Japanese scientist's words.

"These loops include melting permafrost, which releases more methane as it melts. This gas absorbs more heat, which melts more permafrost, which releases more methane, which absorbs more heat, and so on."

"Another feedback loop is the increased heat absorbed by the arctic when ice no longer reflects sunlight and the dark ocean absorbs it. This heat then melts more ice, which reveals more ocean, which absorbs more heat, and the cycle continues. According to James Lovelock, the extra heat that will be absorbed when the floating Arctic ice cap is completely gone is nearly 70 percent of the heating caused by all of the carbon dioxide pollution now present. That, gentlemen, is a lot of heat."

An American spoke. It was one of Professor Henley's students. "Feedback loops include the destruction of rain forests, which release massive amounts of carbon (as much as all the automobiles of the world) because the organic matter decays. This carbon warms the planet, which causes more rain forest destruction in a vicious circle. Then of course there are the methane hydrates (huge deposits of frozen methane) from the ocean floor which, if released will cause an immediate, devastating warming and cause even more methane to be released and then more warming."

All of these processes become worse as the climate warms, he warned, and in turn make the climate more extreme. As they kick in, changes happen faster, then faster again, like a train running away downhill with no brakes. Ladies and gentlemen, we are on a very slippery slope, and it is getting steeper."

"Bullshit!" screamed the denier. "Cap and trade destroys our freedom; freedom of individual choice, freedom to be independent, freedom to live as one chooses. This country was founded on freedom. Al Gore is a jerk."

"Who let him in?" said the speaker, glaring at the middle aged, bald man who had grabbed the microphone.

"Your freedom cannot impinge on my freedom!" yelled a woman in the front row. "Freedom comes with responsibility and we are all responsible for a lot more people than there were in 1776. Your carbon emissions are killing people and you aren't free to do that."

But Stormy had an answer to his ranting. Slowly, but loudly she began chanting, dragging out each letter:

M - I - C- K - E – Y   M - O - U - S - E.

As the crowd began to twitter, someone across the room picked up cadence and soon the whole audience was shouting MICKEY MOUSE over the beleaguered denier who, Stormy learned later, had been paid by oil companies to disrupt the conference.

"The ice feedback loop is obviously the most important!" exclaimed the delegate from Norway. "It's happening now and the entire Arctic ice cap will be gone in a few decades, completely gone!"

"No, the loss of rain forests has a bigger impact," interrupted the Brazilian delegate. "The Amazon forest is one of the world's biggest storehouses of carbon: It is drying up and releasing it to the atmosphere."

"We can't overlook melting methane," said an American. "If the undersea deposits are ever released it will send us back to a climate that was more suitable for dinosaurs."

"Loss of carbon from soils is bigger!" cried Abu, from South Africa. "Soil has been a carbon sink, absorbing fossil fuel emissions for over a century, but as temperatures warm, it is releasing it," he said as he waved his arms to simulate gas coming up from the ground.

"The same thing can happen in the oceans!" yelled a bearded oceanographer. "What has been absorbing a third or more of our carbon emissions is becoming unable to absorb more, and may soon be releasing it. When that happens, civilization is truly doomed."

"Put a price on greenhouse gases!" screamed a very young woman in a red and black dress and a "STOP THE CARBON EMISSIONS" button on her right breast.

The crowd was becoming restless, and as if to confirm their fears, a bolt of lightning struck nearby.

"People. People are the biggest feedback loop," she continued, standing on a chair to make herself heard. "The hotter it gets and the higher the humidity, the more fossil fuels they will burn to keep cool, build seawalls and continue our profligate lifestyle. And don't forget, there are ten thousand billion tons of carbon in fossil fuels, ready and willing to be converted into a greenhouse gas. We are digging it up as fast as we can, building more cars, more coal fired power plants, and more things to use electricity. Like the cartoon character Pogo said–we have met the enemy, and he is us."

"Gentlemen, let's face it, many positive feedback loops are happening all at once, reinforcing themselves and pushing our ability to control climate beyond our reach," the moderator commented. "The combination of methane from melting permafrost, carbon emissions from burning rain forests, less carbon absorption in acidified oceans and carbon released from soil is overwhelming natural climate processes. And if the methane hydrates at the bottom of the ocean ever start melting, we're cooked. Changes that took centuries to happen in the past might occur in decades, or, heaven forbid, even faster."

"Cooked is not a very scientific description," a colleague reminded him.

"True. But that's what we will be; and people need to realize how serious the situation really has become. We have to get away from our jargon and speak the language of the people."

"I agree sir. You are absolutely right."

As the conference ended, an Australian scientist reminded them:

"Because we can't remove carbon that is now in the air, we have committed the Earth to warming three to six degrees, many feet of sea level rise, and massive changes in the systems that provide us food and water. Feedback loops mean the climate will change on its own, regardless of what we do."

"What will it take to make enough people change?" wrote Stormy in her report. "How bad must it become?"

Predictions

The answers, she speculated, were multiple hurricane strikes that made large areas uninhabitable, permanent droughts, food shortages, blackouts, and global epidemics of climate related diseases such as dengue fever for which there is no vaccine. At first, she wrote, climate disasters will affect only thousands of people. But as the storms become more intense, frequent and worldwide, hundreds of thousands, then millions of people will be caught up in its wrath. Like the financial crisis of the first decade, no one will be immune.

Her predictions, and those of the scientists, were to come true, but much faster than they had imagined.

Good Morning

Isolated from the normal world, life on a college campus was good. An 8:00 am class got Stormy up early, and there was always more hot water than anyone could use as she showered and quickly ran downstairs to for the dining room breakfast.

"Hi!" she shouted to the cook, grabbing eggs, and a side of bacon. "Will there be steak tonight?"

"Oh yeah," he replied, "every Wednesday."

"Oh, I think I better have the fish, but save me a glass of that French wine," she replied.

In America, food was still abundant, but that was about to change.

Finishing breakfast, she hopped onto her motor scooter and zoomed away to her field class held off campus at Parch's wetland. Stopping for gas, she was happy to pay only fifteen dollars for the gallon needed to fill up the little machine.

Along the way, she could not help but notice the large number of houses for rent, vacant storefronts, and trash in the street. Even some of the stoplights were not working. *You would think that with 20 percent unemployment, someone would want the job of fixing them*, she thought to herself. But then she recalled how all the bond measures had failed to pass an angry electorate and how much money was being spent on repairing storm damage.

Detouring around the bridge that had washed out in the big flood, Stormy remembered how the most recent 100-year storm had knocked electrical power over a wide area and how looters had taken advantage of it. The power had been restored, but the bridge had not. Part of a larger system, which had spawned five hundred tornadoes over seven states that weekend, such events were becoming all too common.

"Professor Henley, you're early."

He was bent over, examining a line of plastic trash twenty yards from the stream.

"Look here!" he exclaimed. "That last storm brought water way up here. It must be three feet higher than the old flood plain. That explains why that new shopping center flooded."

"And the new housing development too," she replied. "So much damage from one little creek; but I heard it was much worse downstream. Especially for those trapped in their cars."

"These intense rainstorms are popping up all over." said the professor. "But the rest of the class is here. On with the lesson."

"This *is* the lesson," said Stormy to herself.

Dr. Henley Lectures:
"Two of the greatest forces on earth are opposed," said Dr. Henley, dressed in his casual sport jacket and standing at the lectern in the cavernous lecture hall. "These are humanity's insatiable appetite for enormous amounts of low cost energy; and climate change, which is the inescapable result of this appetite."

"Energy on demand, be it electricity, gasoline, heat, or transportation, has grown to be a necessity for our daily way of life," he continued. "Look around you. How would your life be different if you could not turn on a light switch, cook your food, travel miles in minutes, or heat and cool your dwelling? How much commerce would occur without oil, the internet, or electric based communication? Indeed, the world as we know it would shut down. Food would become scarce, faucets may not flow, and without enormous quantities of cheap energy, large centers of population such as Tokyo, Delhi, and Los Angeles would collapse. Because renewable energy now comprises only a tiny amount of the energy that is needed, these things are provided to us from the burning of fossil fuels."

Stormy squirmed in her seat as she listened to his words, scribbled notes and watched in amazement at the climate balance graphic.

"But the more these fuels are burned, the more they change the climate. A great battle has been engaged between these two forces. Incredibly strong storms tear at our power lines, floods destroy our landscapes, drought kills our crops, and sea level rise threatens the homes of millions. As of this day, unlimited use of energy is winning the battles for our souls. But as it does, the seeds of unstoppable climate misery are nurtured. Because carbon, the by-product of fossil fuel energy, cannot be economically removed from the atmosphere, its intensity grows, feeding ever stronger feedback loops that further enrage the climate's wrath."

OMG, my name is based on destruction, thought Stormy, wearing jeans and sitting in the first row. Dr. Henley's predictions bothered her but the fighting spirit in her kept saying that there must be a way to stop it.

"How?" She thought to herself. *"How can this great civilization stop burning fossil fuels, regenerate the planet's forests, and tip the scales in the battle between energy use and climate change?"*

Burning Food is Stupid

Later that night she flipped off her shoes and climbed into bed with her computer tablet. It had been a long day but there was a paper due in class tomorrow. Ignoring the fatigue, she rationalized that if ten percent of the world could go to bed hungry, she should certainly be able to sit in a comfortable room and write.

> *"Burning food as fuel, such as bio-diesel and corn ethanol, is stupid because it takes food from world supplies. Filling one SUV tank with corn ethanol takes enough grain to feed one person for a whole year."*

She had researched the growing food riots in India where drought had tightened its grip. Stretching out on her stomach, she remembered that Indonesian palm oil was still being shipped to Europe to make bio-diesel fuel rather than being used for food. Loud music from the next room reminded her of the riots in Africa because shipments of corn from the United States had slowed because more and more of it was being converted into alcohol to power cars. Rain pelting her window brought thoughts of how the monsoon rains came at the wrong time, shifted off track by global changes in wind patterns, and decimating farms in Africa. Looking at her roommates' cactus, she wrote that the deserts were continuing to expand across the continents because extra heat in the atmosphere had disrupted the planets distribution of rainfall, moving it northward in the northern hemisphere.

She turned to a study by a prestigious think tank and read their prediction:

> *"One third of the planet's land mass will be desert by 2100."*

Wincing, she added that to the summary on her last page. In her chapter on famine, she wrote:

> *"There have been food riots before, in 2007 and 2008 when food prices had doubled, but now in 2022, it is much worse and prices are four times higher. In third world countries, hundreds of thousands of people have banded together, roaming cities, and "liberating" food from the wealthy. Before, the strife was seasonal. Now, it is becoming*

continuous. New Delhi has been wrecked by protesters; rioters tore through the capital in Mexico; students are fighting with military patrols in Iraq and Brazil. Farmers in Australia committed suicide or moved to cities when their crops withered and died. Only in the United States, Canada, and northern Europe do people not have to worry about having enough to eat."

"What is it like to be hungry all the time?" Stormy asked. "What do people do when they can't get enough food?"

"They start wars," her dad said. "So would you if you were starving."

Wealthy people in the United States did not feel the hunger, but when beef prices reached a record high, people began poaching wild bison for food. Sensing an opportunity, Sarpin and the now dominant Republican Party opened up the wildlife refuges and national parks to hunting. Soon hunters in SUV's were blazing away at bison, deer, and anything else that moved.

"Why not," Sarpin proclaimed? "We already use these lands for timber, oil, and water. Why not food? People are hungry."

Farming continued in California's San Joaquin Valley and elsewhere, but farmers were shifting to crops that could withstand the rising temperatures. Wineries moved to Oregon and Washington. In New England, maple syrup production shifted to southern Canada as maple trees succumbed to a changing climate. For farmers, it was a time of change. Those that could not change, went out of business, including Stormy's uncle.

"I give up," Uncle Fred wrote to her. "I can't grow soybeans in Alabama anymore so I might as well retire. I sold the farm and I want you to have this money for college."

Her eyes grew big as she looked at the check for $50,000. Now she could pay off her student loans and continue her studies.

<u>Graduation:</u> "This is Just the Beginning" was the title of Stormy's senior paper. In it she described how climate disruption will come faster and faster as the decades go by.

> *"It's like a slippery slope that keeps getting steeper," she wrote. "Already it looks to be uncontrollable as politicians continue to compromise. Greenhouse gas emissions are continuing to rise dramatically and several scientists have said it is unstoppable, irreversible and happening faster than ever before."*

> *"We are seeing frequent temperatures and rainfall amounts that usually happen only once every 100 years," she wrote. "But remember, climate change is just beginning and it will get worse, much worse. When it does, we will see heat waves and floods bigger than ever occurred in all of human history! The carbon is in the air, and the writing is on the wall.*

> *For 10,000 years, our planetary thermostat has been set at 57 degrees," continued her report. "But when the earth was two degrees warmer, sea levels were 20 feet higher, and the last time it was three degrees warmer, sea levels were 80 feet higher. It may take a long time to get there, but we can't ever lower sea level, nor can we cool the earth. These are things far beyond our technology. We are on a one way street toward sea level rise, disruption of agriculture and dangerous weather."*

"It will get a lot worse," wrote Professor Henley as he graded her paper.

At Stormy's graduation party, the argument became heated.

Playing the devil's advocate, Dr. Henley challenged them to prove that climate change was real.

"Okay," said his precocious young student, "just ask yourself four questions."

"Four questions?"

"Yep. First, is there carbon dioxide in the atmosphere? Second, is the level of it increasing? Third, does carbon in the air make it warmer in sunlight? Fourth, are people increasing the amount of carbon in the air? The answer to all of these is yes. If you answer no to any of them, you better declare yourself God because you just changed the laws of physics and ignored a century of scientific measurements."

"I don't believe any of it," said a drunken student.

"What? If there isn't carbon dioxide in the air, what is that bubbling in your ginger ale and what are you exhaling?"

"Well, okay, maybe there is a little bit of it out there," he stammered. "But the increase is natural."

"Crap!" exclaimed Stormy. "It's rising faster now than ever before in the history of our planet."

"Yeah, but people aren't doing it."

"Right, and I suppose that coal fired power plants and automobiles don't exist either." Stormy was fuming.

"So just how does the carbon in the air make it hotter, smarty pants?"

"Certain wave lengths of infra-red sunlight that are reflected from the Earth can't penetrate carbon dioxide, so they bounce back and create heat. Infra-red does create heat you know. Here, put your hand under this warming light by the food tray.”

“Ouch! That’s too hot.”

"Bravo!" exclaimed Dr. Henley as he raised his glass to hers in a toast.

“Carbon dioxide is our thermostat, and we are turning it up,” she continued, a gin and tonic in her hand. “But we can't turn the thermostat down.”

“So, we'll just take the carbon out of the air,” said an inebriated denier standing next to her.

“You can't remove carbon from the air,” piped up Dr. Henley, who was drinking his usual Irish Coffee.

“Why not?” said the drunk.

“Because it stays in the air for an average of 100 years and there is no politically or economically viable way to remove it,” the professor replied. “Not only that, carbon levels have been going up every year, not down. We're already past the tipping point, so drink up.”

“Tipping point? Is that when you're tipsy?” he giggled.

“No.” said Stormy. “A tipping point is like when you were paddling that canoe last summer: It was leaning, and you moved an extra inch. Remember?”

“Oh yeah, one more inch and over she went. That really was a tipping point.”

“You can paddle along for a long time leaning over, but one little bit more, and over you go,” said Stormy as she almost fell over.

"Unlike your canoe, we can't turn the planet back over," lectured Dr. Henley. "Once you're past the earth's tipping points, climate change is completely irreversible. So pass me another Irish coffee."

"The problem is that greenhouse gas emissions are not like any other pollutant. You know, there was DDT, which nearly drove several species to extinction and even got into people's breast milk. We found substitutes and convinced most countries to stop making it: Problem solved. Then there were the chloro flouro something (I never could pronounce that) chemicals that ate up the ozone layer, which was filtering out the sun's rays so we didn't get skin cancer. Once the world stopped making them, the ozone layer started to heal itself. It took an international treaty, but most countries signed on: Problem solved. But greenhouse gases are everywhere. One third of them come from transportation (cars and trucks, and you know, we all drive those). Most of them come from burning fossils fuels to keep us warm, run air conditioners, and turn on lights. A whole lot more carbon gets into the air from burning forests, and even from making cement! How are we going get people to stop doing all those things?" worried Stormy.

"They won't" replied Dr. Henley. "And that's why the planet is doomed."

A Job and a Story

Garbage men had just hauled away the last graduation party banners and sodden cake when the text message came. It was the *Tribune*, offering Stormy a job at WGMedia in Chicago. Her dream had come true. Quickly packing her new Bachelor of Science in Climate Change degree, she left the hallowed halls of ivy and drove the old sedan with a bumper sticker that read CARBON DIOXIDE MAKES ME HOT down I-80 to the windy city where she parked by the Fullerton Avenue beach at Lake Michigan. As she walked barefoot in the sand, a stranger passed by and smiled. It was George, her future husband, but their rendezvous was to be many years later.

A Mentor

Boothby had worked at WGMedia for years, more years than he wanted to remember. Older, wiser, and a bit of lush, he had tutored new reporters for decades with his compassion, friendly smile, and wide-ranging knowledge of the real world. Boothby liked people, and people liked him. Originally from California, he drove an old VW bus with TAX CARBON, NOT INCOME, and THE AIR IS NOT A SEWER bumper stickers. Recognizing Stormy as a young writer just out of college, he quickly became her mentor and confidant. As for Stormy, she liked his twinkly eyes and rumpled shirts.

A Sage

Boothby was more than a newsman. He had grown up in a family steeped in science, and prided himself on understanding the background of what he wrote. When he realized that Stormy was going to be the WGMedia climate reporter, he wanted to make sure she knew what she was writing about.

"So, Stormy, when did civilization first realize that carbon in the air can trap heat?" he asked.

"In the mid-1880s, when good old Svante Arrhenius first discovered it. But way back then, he thought it would be a good thing, making Sweden warmer. Of course lots of other people have repeated his experiments and got the same results. You would have to change the laws of physics to disagree with the fact that carbon in the air traps heat. And Dr. Henley told congress about it in 1988. They listened at first, but then those merchants of doubt said that the facts weren't true. And you know what, some of those supposedly smart congress people believed the deniers."

"Very good," said the Booth. "Hmmmm, let's try this one. How do we know that carbon levels in the air are increasing?"

"Oh that's easy. Not only do we have the daily data from Mr. Keeling in

Hawaii going back to 1958, there are now satellites overhead that take hundreds of readings every day. Not to mention the gas bubbles trapped in icecaps; those go back hundreds of thousands of years and the data from deep, ancient ocean sediments, tree rings, coral reefs laid down hundreds of years ago and cave stalagmites. The numbers from all these sources match up, and they all say that carbon levels are going up. Haven't you read *The Two Mile Time Machine*? It's all about ice cores two miles deep; they hold a record of how much carbon was in the air more than 10,000 years ago. The ice cores in Antarctica go back even further. All these numbers tell us that the carbon levels in the air we are breathing right now are higher than at any time in the last 15 - 20 million years! But if you don't believe all the scientists, you can buy a carbon dioxide meter and see how much is in the air for yourself. It's a really cool tool. Here, see, I've got one in my purse."

"It looks like an alcohol meter," said Boothby as he blew into it and watched as the carbon level jump from 450 to 1500. "Wow, I didn't know I exhaled so much carbon. Hmmm, you seem to have mastered that one; I don't know anyone else who carries a carbon meter in her purse. But how do we know that the earth's average temperature is really rising?"

"Oh that's easy. Boothby, there are 17,000 temperature monitoring stations. Some of them go back to about 1850. They are on ships out on the ocean, on buoys in the sea, on land, almost everywhere. And where people aren't reading a thermometer every day, ten satellites are sending back data all the time from everywhere on earth. The average temperature has gotten warmer, by about 2 degrees since 1870, no doubt about it."

"Two degrees! That's all! What difference does that make?" Boothby was playing the devil's advocate.

"Okay, the difference between an ice age and now is a really big difference. Right?"

"Right."

"Well, during the last ice age, Chicago, New York, and London were under twenty feet of ice, and the average temperature of the planet was only ten degrees colder than now. So two degrees represents about 20 percent of the difference between an ice age and now. The problem is we are looking at 5 - 10 degrees of *warming* in the few decades. So if ten degrees colder is an ice age, imagine what 5 - 10 degrees *hotter* will be. Yeah, two degrees is a really big deal if you're the climate. It would mean the whole planet would change more than it has since civilization developed."

"You mean like sea level change and all that?" asked her boss.

"The last time the earth was 2 degrees warmer, sea level was 20 feet higher."

"But that took thousands of years to happen."

"Right, but we have been adding carbon to the air 20,000 times faster than it was ever added before! We're really pushing the system out of its boundaries so it won't take thousands of years this time. We don't know how long it will take to rise 20 feet, but it's a one way trip; you can't just push a button and lower sea level, you know."

"So, my young reporter, how do we know that humans are the ones causing the increase in carbon and temperature?"

"Hey Boothby, wake up. We're adding more than ten billion tons of carbon into the air every year. That means every year we burn fossil fuels that took more than 400 years to accumulate. Now that's really screwing up the geological cycle! Hundreds of scientists in dozens of countries have analyzed how much warmer the earth should be based on solar cycles and everything else. You know, the cycles that guy named Malankovich worked out. It's been proven, unless you are a stupid climate denier, that the natural cycles are being completely overwhelmed by all the coal, oil, and natural gas we are burning. Disagree with that and I'll question why you are disagreeing with people who have spent their whole lives studying it."

"But what about the Intergovernmental Panel on Climate Change?" asked Boothby.

"Do you mean the IPCC?"

"Yeah. They said we have several years before it gets really bad, and made lots of recommendations on how to fix it."

"The IPCC summaries are fairy tales," Stormy blurted out. "Their predictions don't even include most of the positive feedback loops, the data they use in the reports are five years out of date, and every sentence of the summary must be approved by every government, over a hundred of them. Do you really think the oil producing countries would allow strong language to reduce oil consumption in the final version of the document, which is usually debated at 4 am in the morning?"

"Well, uh, no."

"The real truth is what has been happening outside the windows of the world for the last forty years. It's much worse than the IPCC said it would be. Their reports are way too conservative."

"Stormy, you pass," said her boss. "Your job now is to convince our readers that climate de-stabilization is real, and that it can really hurt us. It won't much matter for guys my age, but if your generation has children, they are the ones who will suffer, and suffer badly."

George

Politics. That's how I can change the world, thought George to himself as he entered the race for the House seat from Chicago. Choosing his staff carefully and avoiding the bribes, corruption and pay offs that had defined his opponent, he began to craft a career that would last a lifetime.

It doesn't have to be the way it was he told himself in a mirror as he prepared a speech for delivery at Soldiers Field. Politics is a renewable resource, and I'm the one to renew it.

His message fell on fertile ground and within the year, George found himself walking the halls of Congress. But as time wore on, his ambitions grew.

The Subway Floods

Stormy's first assignment took her east, to New York City where for years the subway system had been battling rising seawater and increasing rainfall. There had been several near misses, such as in 2004 when hurricane Frances inundated the city, in 2007 when disaster had been averted only by calling in huge water pumps to keep the tunnels dry, in 2011 when hurricane Irene raised tides to within an inch of flooding the underground train tracks, and in 2012 when hurricane Sandy actually flooded parts of the system. Indeed, 15 hurricanes and major storms had hit the city since 1815, and much of the subway system was just above (or below) sea level.

Standing in the dimly lit station, Stormy felt vulnerable when she realized she was 14 feet below where the ocean would be were it not for concrete walls and pipes. The water, she hoped, would stay away. Commuters rushed by her, still dripping from the downpour they had escaped above where fetid street runoff began to swirl around clogged storm drains. At many of the curb inlets, discarded plastic bags, newspaper, beverage bottles, and tree leaves blocked the water from its humanly intended path, sending it further downhill, seeking out the lowest place where Stormy stood, reading the transit map.

Waiting for a train and shaking water from her windblown umbrella, she was first to notice the miniature waterfalls, surging down the steps, past the tollgates, and across the floor of the station.

"Oh shit, it's really happening," she thought to herself. *"I've got to get out of here."*

But the 5:00 commuter rush was on. Hundreds of people pushed past her to catch the incoming train and she was barely able to make headway toward to exit.

"Get back out!" she extorted them. "It's going to flood."

Grim stares were their reply, as if she were some strange freak babbling about a flood. This was New York. Such things did not happen here.

At the station entrance a child asked, "Mommy, why is that pipe bubbling?"

"Come away junior," said his mother. Afraid of the unknown, she could not know that the submerged pipe, designed to vent air from the subway, was now gushing water into the transit tube below.

Shoppers on the streets of New York saw water cascading across sidewalks and down stairs leading to the subway. Water gushed past storm drain inlets as they were overwhelmed, intersections became glistening pools eight inches deep as the drenching grew more intense. For three hours the downpour continued as the sputtering hurricane stalled, searching for a new direction. Transit workers brought in four emergency pumps, but one was knocked out by lightning and another failed from perpetual lack of maintenance. The remaining two pumps were not up to the task.

The massive storm hit right at high tide, dropping more than four inches of rain an hour onto a storm drain system designed to handle only an inch and a half. Upgrades had been proposed, but were never funded after the mortgage meltdown of 2008. There was no stopping the water this time as it surged in from street vents, stairwells, and utility pipes, flooding the soon to be abandoned tunnels.

Astounded passengers in underground stations waded through three inches of water on the platforms, watching in horror as the oily concoction began to rise next to the metal rails below. As the water bridged the electric rail carrying 600 volts, sparks flew, trains stopped, and the lights went out.

Screams filled the air until the dim glow of emergency lights illuminated the flooded station and panicked people surged for exits. Emptied of human presence, the underground tunnels assumed an eerie look as water continued to rise, carrying coffee cups, newspapers, and a forgotten briefcase. Rafts of the combustible litter floated down the train tunnel, and then caught fire as sparks from the third rail flashed in the darkness. The flickering fires cast an eerie glow against dark tunnel walls as the burning trash islands drifted aimlessly with the currents.

Standing on a step just above the lapping water, Stormy used her cell phone to record the disaster, sending the images out to an unbelieving world while interviewing shaken passengers.

"Have you seen my little boy? That mob pushed him away. Please help me find him." Water dripped from her hair but Stormy could tell she was crying. A team of young basketball players agreed to help, fanning out on the street above, and calling his name.

Stranded, not knowing where to go, hundreds sought shelter in local stores. "Oh my God, how am I going to get home?" exclaimed a distraught office worker. "It's those damned terrorists," muttered another. "It's our stupid mayor," was another explanation. "An act of God," said a housewife.

Fifteen minutes passed; twenty, then thirty. At last the lost boy was found and re-united with his mother. But he would live to see another, final subway flood in 2050.

On her fifteenth interview, Stormy found someone who realized that this had been predicted: Predicted by climate models, transit planners, and even his uncle Harry who ran the newsstand. But the flooding had not been predicted by weather forecasters. Just as it happened when Stormy was born, their outdated computer models, designed for an earlier climate, failed to foresee the intensity of the storm and the city was caught unprepared. Dozens died from electrocution when the third rail became flooded, electrifying those trapped in stranded rail cars. It would be many weeks before the subway system was cleared of silt and re-opened at a cost of several million dollars. Economists called it the Great Storm of 2028 and estimated the loss to Wall Street in the billions.

Conversations in a Pub

Huge raindrops pelted them as Stormy and Boothby ducked into their favorite pub and sat down by a fiddle player who was tuning his instrument. The Tipsy House boys were playing and the dancers were moving to the reels, jigs, and waltzes. Irish flags and posters of patriots lined the walls and kegs supported a bench by the stage.

The seisun was in full swing and for Stormy, it was a sign that all was right in the world. A sign that people could take time away from working to indulge in music, that universal language of brotherhood, life, and happiness. When she heard the notes, Stormy always felt relaxed and content. Free from necessity, her mind could wander, create ideas, and become open to those around her. It was a magical experience that drew her toward it without resistance, so she let herself be taken as she gave her mind to it. Space faded away as the melody prevailed. Tomorrow could wait as the 4:4 beat carried her away. Time didn't matter. The harmony was omnipresent. Life's cares and worried faded because tonight the music was playing. It was only for a few hours, but what else could put the world's problems at bay?

For Stormy and Boothby, the pub was a refuge, an escape from the increasingly bad news that dominated the media. Sitting in a dim corner, they ordered dark, draft beer and began discussing the New York story.

"Focus on the human angle, but don't leave out the climate factor," cautioned Boothby. "Climate affects everything people do."

"That's true," murmured Stormy, brushing her long hair away from the drink. "Everything we do."

Boothby continued. "The problem is that politics operate on a short term cycle, election to election, but global warming is long term, many centuries really. If we try to stop it two years at a time, we will fail; and fail miserably."

"The other problems are the taboos," she replied.

"What taboos?" replied Boothby.

"Population growth and political crap: Hardly anyone will admit that too many people on the planet is the elephant in the room, and there is one particular political party in this country is responsible for preventing the United States from doing anything significant about climate change. How can you solve a problem when people are afraid to talk about it?"

"Hmmmm. Given the nature of human nature, it would seem that climate change is going to win," said the Booth. "People need to work together to solve the problem, and they can't."

"But you know, it's really wealth, not the number of people in the world that is causing climate change," continued Stormy.

"Why so?"

"Because population growth is coming from the third world, the poor countries, and those people hardly create any greenhouse gases. The increase in carbon emissions is coming from coal fired power plants, cars, trucks, airplanes, monstrous houses, yachts, and anything we want flown in from half way around the world. If you can afford it, it's okay to belch out greenhouse gases, and the wealthy are doing it."

"I don't believe in this climate stuff," said the bartender as he delivered their drinks. "The government reports say it could be bad, but they're not sure."

"We're going to *not sure* ourselves into oblivion," replied Boothby.

"Nah, in six months everything will be back to normal."

"It can't be done," said Stormy. "The people on this planet cannot reduce carbon emissions enough to avoid major climate change. Prove me wrong. Do it."

"Back to normal? If you believe that, I've got a bridge in San Francisco I'd like to sell you," replied the sage. These optimists were getting on his nerves. "Maybe I'm just being a pessimist," he said.

"No, I don't think so," said Stormy. "Just because most of the problems are happening in the poor countries doesn't mean they won't happen here."

"Yeah," said a drinker at the bar. "Just ask the people who used to ride the New York subway before it flooded."

After beating her at pool, Boothby offered to buy Stormy a drink. Thankful for a chance to bend his ear, she asked how he thought this climate change thing would end.

"Every system will be affected," he replied: "Every system on earth."

"Oh bull, how can that be true?" she said in a sudden and involuntary manner.

"Well, consider cities: When hurricanes hit Miami, New Orleans, Houston, and yes, even Washington D.C., and when they hit several times in ten years, large parts of those areas will be abandoned, permanently."

"Permanently? Why?"

"Because people wanting to rebuild will not be able to get insurance, and without insurance, nothing gets built. The insurance companies are already bailing out."

"Oh, that's true, that's true," she said. "But where will those people go when their homes are destroyed and they can't re-build?"

"Some will become climate refugees, and unless they have more money than most people, some will end up living in slums."

"Well, how about ... hmmm, transportation?" asked the cub reporter.

"What happens when hundreds of bridges are washed out by floods the size of which we have never seen before, when ports are damaged by sea level rise, and when coastal airports are under water? That stuff will cost money, a whole lot of money and people are not going to want spend that money when they need it for their own problems or for retirement. The baby boom is retired you know. Eventually, the New York subway system will end up as a river because it is below sea level and the bond measure to fix it may not pass."

"Oh Boothby, enough: Get me another drink. I don't want to think about it."

"But then the big ones, the ones that will really get us, will hit," said the Booth.

"Good Lord, what are those?" she asked.

"Food and water: Oh sure, there will be wars over oil first, but eventually it will come down to not enough food and not enough water; basic survival."

"Agriculture is acutely sensitive to climate," continued Boothby. "World food supplies are already too small to feed everyone. The more the climate zones shift, the hungrier people will get. Not only that, heat waves and drought will damage the soil, and you can't feed a hungry nation without good soil."

"Won't the farmers just move north?" asked Stormy.

"You can't grow grain in Canadian swamps," replied Boothby. "Most of the arctic doesn't have enough soil."

"So what would it take for the world to stop climate change in its tracks and avoid all this damage?" muttered a pensive Stormy.

"Worldwide cooperation on a scale that we have never seen," said Boothby, staring off into space.

"Bigger than what we did in World War II to stop fascism?" asked Stormy, looking up at him.

"Oh yeah, and bigger than the space program too," responded Boothby.

"That would make this the biggest challenge that mankind has ever faced, something that has never happened before."

"Given our history, such changes in human response will take too long," said Boothby, his voice rising in a warning tone.

"Do you mean like the changes we had to make when scientists realized that we were not the center of the universe, or when slavery was outlawed?" asked Stormy.

"Exactly," replied Boothby. "Those types of changes require shifts in religion, personal views of the world and governmental responses. But in this case, it needs to happen mighty quick."

"I just hope we are up to it," whispered Stormy.

An Irish jig played merrily in the background as they stared into their drinks. It was music honed by centuries of players, but Stormy wondered how long it would last if civilization declined as Boothby predicted.

<u>An Arctic Incident</u>

It was lonely out on the little research vessel as it moved slowly north of the Beaufort Sea. A few polar bears, sea birds, the northern lights, and a beluga whale made their appearance, then disappeared. So it seemed strange when a bright light appeared in the north, coming closer and heading directly toward them. But this northern light had guns, and people in military uniforms. Quickly, they overwhelmed the American geologists, charged them with trespass in Soviet waters, and hauled them off to Moscow. Russia wasn't kidding. Arctic oil was theirs and they were willing to fight for it. Their country had already invested in forty ice-resistant oil platforms, 60 ice-resistant tankers, and twenty natural gas carriers.

In Washington, Sarpin was furious and the national response bordered on full-scale war. But her talk was not as strong as in the bars and taverns of rural America. Times were tough and people needed someone to hate. Nuke 'em was one response and "Drill Baby Drill" once again became a rallying cry. A very old United States coast guard cutter was dispatched, but by the time it became trapped in the Bering Strait when pack ice shifted, cooler heads prevailed and the Soviets backed down. The cutter had to be rescued by the Soviet fleet of ice breakers, which greatly outnumbered the aging American ships, a story that Boothby wisely buried at the bottom of page 27.

A commission on Arctic territorial rights was established, but it was to meet for many years without reaching a consensus. China, sensing an opportunity to obtain oil for its rapidly growing fleet of automobiles, sided with the Soviets while Japan tried to remain neutral. Russia claimed the entire arctic and Norway, Canada, Sweden, Finland, Iceland, and Denmark (which owns Greenland) all claimed areas that were claimed by others. As the talks dragged on, the Norwegians continued to build an Arctic Navy, South Korea began building ice breakers, Japan began tapping gas hydrates in the Mackenzie Delta, and Chinese "research" ships were increasingly seen near the North Pole, which was now an open ocean all summer.

Assigned to do background research on the issue, what Stormy learned was troubling. Gasoline shortages in America were increasing and more than double the oil reserves of Saudi Arabia lay under the Arctic sea, now ice free for most of the summer.

"Double the reserves of Saudi Arabia: That's incredible. Why didn't I know about this?" exclaimed Stormy.

Arctic gas reserves were also huge and rising prices driven by a growing global population had made it profitable to drill. Oil lease sales were going for one, two, three billion dollars and even more. Would society be able to remain civil in the face of such wealth and potential conflicts?

The Climate Editor:
The incident prompted Stormy to demand more from WGMedia.

"There needs to be a daily climate column and I need to write it," she told Boothby.

Boothby smiled. He had been expecting this. "Okay, you shall," he replied. "I appoint you the first editor in charge of lowering carbon levels. Go for it girl. But be aware, we are going to send you around the world for these stories."

"Oh, I think I can handle that," she said. Stormy was in her late twenties, and couldn't wait to make a difference. "If people understand what is happening and what is at risk, they will get these carbon emissions under control. I know it," she told her mom, who replied with only a sad smile.

Who is TERRA?
After two Irish coffees at the pub, Stormy was getting philosophical.

"I know it's just physics, mathematics, and all that, but what would TERRA be like if she walked in the front door?"

"I don't think she would fit into this room," replied Boothby.

"Yeah but if she could talk, what would she say?"

"Get out of my way maybe, or stop feeding me these greenhouse gases because I'm getting fat." Boothby was a bit loose, too. "TERRA really is getting out of control," he continued.

"Out of control: Yes that's true. She's like a robot that does what she's programmed to do," replied Stormy.

"Out of control, and if we're in her way, look out: Floods, fires, drought, and famine." Boothby looked tired.

"Yep, and then we respond to all this climate change with riots, wars, quarantines, and politics. TERRA, she doesn't care. She's neutral: Not Republican, Democrat, or Sinn Fein; not Catholic, Muslim, Lutheran, or Buddhist. Hey Boothby, TERRA is agnostic!"

"Yeah, and she's not American, Chinese, Irish, or South African. I guess she's the original everyman."

"The trouble with people," Stormy continued "is that we think we can control her."

"How can you control a force?" asked Boothby: "A force that changes the world's weather, from tornadoes to lightning to the amount of rain that falls?"

"Yeah, she's powerful all right. But the main thing is she just doesn't care."

"That's a very dangerous combination," said Boothby. "She can do whatever she wants to anybody, anywhere on Earth and she doesn't care. It's like we are invisible to her."

"We really are. If she walked in here right now she would just measure the amount of carbon in the air, figure out how much heat it would trap in sunlight, toss in the physical laws, and mix up a batch of weather. And if that weather floods your home, kills your crops, or melts your water supply, she doesn't care."

"We could gang up on her," said Boothby, as he raised his eyebrows.

"Nah, we haven't been able to agree on anything enough to do that," said a subdued Stormy. "People would rather fight each other than unite in a common cause."

"Then maybe it's our fault, not hers."

"You know, I think you're right. You can't blame a physical force, can you?"

"I can try." replied Boothby. " Wind, it's your fault. Ice, you shouldn't melt. Corn, grow where there is no rain."

Boothby was waving his arms over his head.

"Umm, Boothby, nobody's answering."

"Oh shit, you're right. The laws of physics win. We lose: Bye-bye civilization."

State of the Earth

Stormy's next column summarized the state of the world as it struggled to contain carbon levels.

WORLD COALITION CRITICIZES U.S.: "A world coalition of China and under-developed nations heaped vitriolic criticism on the U.S. for failure to address climate destabilization," she wrote. "Joined by Latin American, Indonesian, and African countries, they have begun a global boycott of U.S. products and services, causing panic in the stock market.

*"American carbon emissions have destabilized the weather"
she quoted an Indonesian representative as saying. "Our
citizens are dying in floods, storms, and drought while you
continue to burn fossil fuels as if we did not exist. Every year
the carbon emission increase and we suffer even more. We will
not tolerate it any longer."*

A reader responded: It was Cedric, her old high school nemesis, who
had become a vice president of the World Oil Company.

"Your analysis is faulty," he wrote. "Carbon levels have not increased
temperatures as much as carbon itself has risen, so carbon can't be the
cause of our recent, temporary warming," proclaimed Cedric. "And don't
forget nine-eleven. Climate change is just a liberal plot to distract us from
terrorism, which is the real threat."

In reply, Stormy wrote: "Do you know what? The average Joe Plumber is
more concerned about the price of food than nine- eleven. The tea party
types tried to have a 25th anniversary rally about the planes that flew
into the old World Trade Center, but hardly anybody came. That's not
surprising: One person in four wasn't born when that happened. They said
they couldn't afford a decent meal and that was more important than some
terrorist in some other country whose name they couldn't pronounce."

"In reality, what is now happening in front of our eyes is that temperature
is racing to catch up with the massive amounts of greenhouse gases that
we have put into the air." The reason that we are not burning up now is
that there is a lag time because the ocean has absorbed most of the carbon.
But it cannot continue to absorb everything we throw at it. The oceans
are becoming more acid and cannot absorb as much carbon. When the
earth's system catches up, it will be a lot hotter."

In a private email she told Cedric to go back to his video games. "This is
the Earth we are talking about, not your private market place."

A Mysterious Stranger

"Boothby; who is that strange guy over there in the corner? He's always alone and he's always reading something on that tablet and playing with a calculator."

"Yeah, and sometimes he scribbles things on little pieces of paper. Let's go find out who he is."

"What, you mean just walk up and say hi?"

"Sure. If you did it, he would think you were hitting on him. But I'll just put on my journalist hat."

Stormy marveled as Boothby, everyone's friend, offered to buy the man a drink, sat down with him, and began a conversation. The old man introduced himself as professor emeritus, college of life, school of hard knocks. But it soon became clear that he possessed an intelligence to match his wit. He had studied broadly, read widely, and written more than a few books.

"Join us!" shouted Boothby over the music from the stage. Stormy walked over.

"Dr. Henley!" exclaimed Stormy. "I took all your classes at the University."

"I remember you," he said. "You were my best student. The only one that realized that global warming was really climate destabilization, more extremes of climate."

"And I remember you," she said. "Your lectures on how much change would be needed to avoid disaster were the best that I ever heard."

"You know, as much as I tried, I could never get it across to the politicians that this was something bigger than the efforts we put forth for World War II or going to the moon," replied Dr. Henley. "When I told them that changing every car on earth to use half the fuel they use now would only remove a third of the emissions needed to stop climate change, they didn't want to believe me. Compromise, that's all they wanted to do. They needed the votes from coal mining states and the money from big oil companies so they tried to pass weak legislation, and in the end, even that failed, blocked by lobbyists—who were paid to do it—and the Sierra Club—which wanted laws so strict that they would never pass."

"I guess the Sierra Club and other super greens would rather have climate change than cap and trade." said Stormy.

"It's really not an either/or choice, but what they are getting is climate change," said Boothby as he nodded in agreement.

"People don't realize how much carbon is going to be released. It really can overwhelm the planet," sighed Dr. Henley.

"Will the Earth become like Venus?" said Stormy.

"I hope not," said Dr. Henley. What we needed to do, ten years ago, was to start phasing out coal plants; replacing them with solar, wind, and geothermal energy. We should have stopped destroying the Amazon and arboreal rain forests and put a tax on the use of carbon fuels, or at least a price on it that could be traded so the market could shift our energy use. It would have been cheaper to pay Brazil and Indonesia to stop burning trees and peat bogs than to build all these sea walls and try to stop flooding, heat waves, and drought."

"We had a chance to shift wealth to southern continents and limit poverty by paying them not to pollute the atmosphere like we did for a century, but we chose not to," Dr. Henley continued. "So now we have poor, radical religious factions throwing bombs at our way of living. In the end, everyone will suffer. Famine and drought don't care who they kill, and our money will only save us for so long."

<u>Uncle Pierre and the Smokestack</u>

"Do you know a Pierre La France?" inquired the voice on the phone?" "Uh, yes, what has he done now?" replied Steve.

"He's been arrested for stuffing a blanket down a smokestack," came the answer. "Said he was going to stop these God damned carbon emissions that killed his wife, or something like that."

"Oh Lord, not again. I'll pay the bail and pick him up."

Pierre was proving to be a problem. Retired and with time on his hands, he had decided to become a one man, super hero to stop carbon emissions that were killing the Earth. Steve had tried to reason with him, but secretly agreed that he was right. Anez worried that he might be losing it. Stormy just sat back and learned from his episodes of protest, arrest, release, and protest.

"He has a just cause," Stormy wrote in her journal. "But Pierre is before his time. Ancestors will recognize him as a prophet."

Released from jail, Pierre was livid. "We need more hurricanes, famine, war, and sea level rise," he told the reporter as he left the downtown police station. "Go storms, go floods, go drought, war, and famine," he said. "They're the only thing that will make these God damned American politicians realize what is happening. You can't pray your way out of a drought. You can't stop climate change with political compromises. You can't ignore irreversible, unstoppable global warming. Christ, the whole Arctic is melting. Wake up!"

Shocked at his ambivalence to the human suffering that he encouraged, reporters asked him if he really wanted millions of people to die from tornadoes, disease, and social disintegration.

"Yes," he replied. "Yes, it will be a trivial number compared to what will happen if we don't reduce carbon emissions. My wife shall not have died in vain. I want her death to be known to the people who keep on burning fossil fuels. They killed her, sure as hell."

The next week he was arrested for painting "Climate Change Train" in huge letters on the side of a train carrying coal to a Montana power plant.

But in a month he was back, surveying where coal trains slowed down for a curve and where bushes could hide someone next to the tracks.

TERRA - Gaining Strength

Major but seemingly isolated climate catastrophes marked TERRA's actions during the third decade, but as the sun rose at the end of 2029, billions of tons of greenhouse gas per year were still pouring into the air. It was true that there were more solar panels, an occasional wind generator, and more electric cars, but the earth's eight billion people were hungry for energy, and most still didn't care where it came from. New coal-fired power plants continued to be built in China; big V-8 engines still powered some people to work, and drilling for oil continued at a frantic pace around the world, wherever it could be found.

TERRA's temperature reached 58.8 degrees Fahrenheit, fueled by ever increasing carbon emissions. Ominously, methane levels were also increasing, dramatically feeding the atmospheric frenzy of heat, rising water vapor, and storms.

Gone were the environmentalist shouts of "350" as carbon levels in the global atmosphere reached 435. Many people and a few governments in more enlightened nations were trying to slow emissions of the gases, but were blocked by those that profited from the burning of coal and gasoline. Some people had given up. But most did not really know what was happening, and did not care. Nor did TERRA care when she unleashed her first hurricane on San Diego, killing thousands of people.

Stormy Dreams - Bewilderment

Just like her mother, Stormy had vivid dreams. As she slept, dark red oceans, millions of flowers, and thin forests passed before her eyes, but they quickly moved away, as if fleeing from the dark clouds. The dancing bright lights frightened her, as if to warn of things to come. Billions of stars flickered above and she remembered that her mother had said they represented the people on earth. But these stars seemed to be getting closer, menacingly closer. Buildings had replaced the jagged mountains. In a particularly real dream, children approached her, and then vanished. Were these to be hers?

In another vivid dream she was fighting, fighting against something black that was all around her. Despite her screaming, it continued to grow, enveloping everything around her. Later, she realized that it was carbon in the air she was trying to fight. Intrigued, she checked what carbon levels had been doing since fluorescent light bulbs, traffic lights powered by diodes, and electric cars had become popular. Relentlessly, they were going up. Humanity's meager efforts were not working, and the blackness was getting closer.

At breakfast, remembering her dream, she wondered what had been trying to communicate with her. Was it the climate?

CHAPTER FIVE– BEYOIND THE TIPPING POINT

2030 - 2040

Atmospheric Carbon Level – 435-460 ppm

Average Earth Temperature -59°
Sea Level Rise - + 15 inches
Earth's Population - 8.0 billion people

Change Becomes Rapid

It has been six decades since carbon levels began rising every more rapidly, and TERRA was responding faster and faster. Storms were more frequent, more widespread, and more intense. Their offspring, misery, sickness, and poverty also grew, spawning refugee crises, political instability, and armed conflict. But civilization as a whole was unable to connect the dots, so population, carbon emissions, and consumerism continued to grow in tandem.

A Changing Scene

Stormy was in the prime of her young life, attractive, witty, and ready to take on any problem. The problems however, were growing. Summer water rationing limited how long she could spend in the shower and a reporter's salary wasn't keeping up with inflation, teetering at fifteen percent. Electricity costs had tripled as air conditioners struggled to cope with heat waves and gasoline prices were abominable.

Walking to work, Stormy gave a gaunt, hungry looking beggar a dollar and wondered at the sign he held: FORMER FARMER - CLIMATE REFUGEE. He was emaciated, sunburned, and looked like someone from a third world country, as he said "you can't grow soybeans in the south anymore. It's just too damn hot. I used my last money for bus fare from Mississippi and I haven't eaten in two days. I lived in a homeless shelter but there were too many people there and not enough food."

His gaze melted her and Stormy gave him ten more dollars.

"Thank you," he responded. I don't drink and I don't do drugs. This will be for food."

Stopping at the convenience store, she glanced at a paper to see if the editor had included her last story. But the headlines were all negative, stressing the rising crime rate, international recession, drastic drop in stock prices, and rising cost of wheat, corn, and soybeans. "We used to be immune to these things," she thought. "But more and more it is one planet, and America is being swept up in world events. Wars in the Middle East and Africa keep draining the treasury. Epidemics of dengue fever, malaria, and crop diseases weaken the economy. Hungry radicals throw bombs. Even the New York Mets are going bankrupt." Arriving at the office, she realized that it was all interconnected, and climate change was linked to all of it.

A Whirlwind Romance

At age thirty, Stormy had seldom been with a man; but she wanted to. Most of the guys she had dated were immature, clumsy, and had their minds mainly on her bed. So when a young Congressman from Illinois appeared at the local pub, singing along with the band, she decided to step out of her shell. Unbuttoning an extra button on her blouse, she sat down right across from him. He could not help but notice, and (given the shy smile and cleavage before him) her youthful appearance was the last thing on his mind. Wise beyond his years, George had made his mark by barging past conventionality, and being a little bit inebriated, he took a chance, asking her to dance. She accepted, but the band began playing a waltz. Holding her at arm's length, they whirled across the floor. Stormy kept from becoming dizzy by focusing on this face, and by the time the music ended, she found herself kissing him. It was a long, lingering kiss and when it ended, she realized the other patrons in the bar were cheering.

They sat down, shared a few drinks, and talked about anything and everything. She wanted to put her arms around him, press him close to her body, and feel his touch. It was to be the first of many satisfying conversations they would have and when the night was over, Stormy knew that there was something else in her life besides a job.

George had never met anybody like her. He was drawn to strong, self-confident, and intelligent women, and he found all of these things in her, and more. Stormy had a keen sense of people, and the ability to put them at ease without being patronizing or overbearing. On top of all that, she was an excellent chef, great in the sack, and a realist. He could not imagine anyone better for him.

But Stormy was not sure. "Why get married when the planet is plunging toward oblivion," she wondered. "What kind of life would our children have?"

George persisted however, and when she found the tiny diamond ring on the tail of a little, stuffed dog he gave her, she said yes.

Two months later they were married.

A Honeymoon in the Arctic

"I've always wanted to see the North Pole," was Stormy's response when George suggested they book passage on the Northern Star cruise boat.

"The North Pole won't be there: It melted last year along with the ice," came the reply from her husband. "We may see narwhal and killer whales, but there won't be any polar bears, seals, or walrus. When the ice melted, they stayed on land, or drowned."

A blank look pervaded Stormy's face.

"Oh my god; where has Santa Claus gone!?" she exclaimed.

"Unless he went to Barrow, I guess he drowned," said George: "Along with the polar bears."

"But what about Christmas?" said Stormy? "Has a century old tradition been destroyed by climate change?"

"It would appear so," said George.

"Will the south pole ice ever melt?" she asked.

"If it does, we won't be around to see it," said her new husband. "Sea level will be up by about 80 feet, probably more. Most coastal cities will be completely under water."

"Oh. But you said narwhal would be gone, too. What are narwhal?"

"Narwhal are small whales with big, long, single tusks. Sailors used to think they were unicorns. They're the most endangered of all the arctic animals but hardly anybody knows about them since there are none in captivity. We'll also see oil drilling rigs."

"Really?"

"Yep, for the first time in millions of years, the Arctic Ocean is ice free; we'll be witness to history. But that means it is open to exploitation, and there are truly massive oil and gas deposits 5,000 feet below surface."

"Wow, and the 8.3 billion people on this planet are desperate for those fossil fuels. We're almost out of oil now and our civilization depends on it. What would we do without it? George, I'm worried about this oil situation."

"Don't report it in your blog yet, but there's a memo circulating in Congress that a Middle East/Soviet coalition has shifted oil exports to Russia, China, and Indonesia, while stopping shipments to the United States. Not only that, Mexican oil wells are going dry."

Stormy frowned as she remembered that although Canada had increased shale oil production, it was being refined in Texas and shipped to Brazil and China. Meanwhile, Phoenix and Las Vegas were bidding against

Chicago and St. Louis for energy to generate electricity for their ever-increasing need for air conditioning.

"Our energy supply is drying up," she said.

"Yeah, and I paid $25/gallon for gas last week," said George.

"We should have bought an electric car."

"We're in for a major economic depression," said Stormy.

"More like a major oil war," said George.

The Earth's Marriage Vows

Thinking back on their marriage vows, Stormy realized that people should devote their lives not just to each other, but also to the environment that made their existence possible. Over a candle-lit dinner one night she wrote what became known as the Earth's marriage vows:

> *We, the people of this Earth, take you our planet, to be our morally wedded partner, to use and to sustain from this day forward, for better or for worse, for richer, for poorer, in sickness and in health, to love and to cherish; from this day forward until death do us part.*

"If everyone took this vow, we would all be a lot better off," said George, "and perhaps TERRA would not be such an angry bitch."

A Son and a Daughter

"It was 2032 and I remember it like yesterday," said George as he recalled the birth of Jack, his son. "Then, two years later, you came along," he told his new daughter.

Born into a rapidly changing world, Jack and Susana were to lead very different lives, but the common thread would be events beyond their control. The history of their people had been one of increasing dominance over the environment, but it was becoming more apparent than ever that limits had been exceeded as people became a geological force. Among these limits were atmospheric carbon levels, which were now increasing by almost 3 parts per million every year.

The little babies could not know this. Nor did anyone know how TERRA would respond to such an immense change, one which was happening at a speed that ridiculed geologic time. Susana giggled a lot, but carbon levels in her world had reached 450, trapping ever more heat in the lower atmosphere. It was a sweltering day in April and soon she was not comfortable enough to giggle anymore.

"Don't worry," George said. "This brown out will be over soon and the air conditioning will come on again."

Conversations in a Pub

It was Saturday night and the joint was jumping. On stage, a young singer was belting out old folk songs:

> "I don't even know where we are,
> They tell me we're circling a star,
> Well I'll take their word, I don't know
> But I'm dizzy so it may be so."
> (Defying Gravity, by Jesse Winchester)

Abruptly stopping, he pointed to Doth the bartender, and asked him if he was dizzy.

"Uh, no," came the startled reply.

"Well do you think we're really circling a star, and that the Earth is a big, round ball, spinning around really fast?"

"Uh, yeah, I guess so."

"But that would make you dizzy. So if you're not dizzy, why do you believe that we're all spinning around in a circle?"

"Uh, I just believe it because I've seen pictures, and everybody tells me that."

"But you can't feel it, right?"

"Uh, no."

"So that tells us that we can't always believe our senses, which are designed by nature to react to things that change much faster than spinning around the sun: Right?"

"Yeah, like when a tiger jumps out at you, you know it."

"So, dear audience, Doth has just explained why we can't really sense climate change all the time; it's happening too slowly for our eyes, ears, fingers, nose, or tongue to believe. Thank you Doth!"

Doth took a bow.

Later, Boothby ordered his usual gin and tonic from the bartender. Slowly sipping on it, he reflected: "We can solve technical issues; like, you know, building things. But we can't solve the social issues. And that is what will prevent us from stopping climate change."

There was a break in the music and Stormy turned to him.

"Whatever do you mean?"

"Except for limiting chemicals that destroy the ozone layer, the people of this planet have never been able to agree on something that is worldwide.

But the ozone problem is baby stuff compared to global warming. Wealthy countries created the problem, but they can't solve it without the poor countries. The poor can't afford to fix it and they want the wealthy nations to pay for it—which, of course, they won't. So nobody agrees on anything, and the problem just keeps on getting worse.

"Consider water," he continued. "It would be easy to build huge dams and reservoirs to catch the runoff from the mountains and then build canals to carry it to the people. The Romans were doing that many years ago until their empire fell apart and people started using aqueduct bricks for houses. That could happen here too."

"With all the warming, there is lot of rain and hardly any snow," Stormy interrupted. "So why can't we just build more dams to catch the rain?"

"Because people have to vote on bonds to build the new dams, and if the bonds don't pass because the people don't want more taxes like Sarpin says, then the dams won't get built. The result is not enough water for everybody. That's the social issue we can't solve. But the bigger problem in a lot of places is just not enough water, like in California. Like you said, the precipitation out there now is mostly rain, not snow. When it was snow, it would slowly melt all summer, but now it just runs off into the ocean and by August, Los Angeles is really short on water; San Francisco, too. So, your next news article will be on the California water wars," said Boothby. "Farmers need the water to grow their crops, cities are drying up, environmentalists are desperately trying to keep water in the rivers for the fish, and all hell is breaking loose. I heard they impeached the governor again."

"Oh good, I get to go to California," said Stormy. "It looks like there is going to be a permanent drought in the southern part of the state, worse than the 1930s dust bowl in the mid-west. What happens then?"

Boothby took a long drink from his glass. "First, parks and lawns start to turn brown. Then, if industry can't get enough water for manufacturing, factories move away, and lot of the people go with them, maybe to

Canada or somewhere where there is enough water. If it gets bad enough, the food supply will start to shrink. It already has in the poorer parts of the world like India, where protein is unavailable in most areas, and in Texas, which is rapidly becoming a desert."

"Yeah, and I read in a blog that a third of the people in Africa, Indonesia, and parts of South America can't afford rice, potatoes, and wheat. But it's not a social issue in those places," said Stormy. "It's because the monsoons are dumping the rain out over the ocean or somewhere else instead of on land where it used to go."

Boothby frowned, thinking of the tens of millions of people who would not have enough to eat today.

"I wish that climate victims could vote in a worldwide election," said Stormy. "That would really turn things around. What do you think Boothby? Are there more climate victims than climate deniers?"

"Millions of people are hungry and homeless due to climate change," said Boothby. "South Pacific islanders, Inuit on the Alaskan coast, people in Indonesian river deltas, not to mention those in New Orleans, Missouri, and Miami whose homes were turned to splinters by hurricanes and tornadoes. It's really sad, immoral really, that the same people who are responsible for their misery are still lobbying Congress to stop any action to reduce carbon levels. If there is a hell, they will surely be in it someday."

"Social, technical, political, whatever: It is really a different world out there. It's changing fast. Nobody has been able to stop it, and you know, soon it will be our turn to be hit by it. But I got to go. See you tomorrow in the office. You've got to finish your stories on the western wildfires for good old WGMedia before you leave."

Dengue Fever

But Stormy's trip to the west was interrupted when she got news from Georgia where Steve and Anez had been camping in the Okefenokee swamp. The trip had started well, with sightings of alligators, ibis, and other exotic birds. On the second day however, a torn mosquito net proved to be their downfall.

"I didn't sleep a wink last night," said Anez. Steve winced as he looked at his wife and saw the dozens of mosquito bites on her arms and face.

"Don't worry," he said, "they got me too, but the itching will go away soon."

"Right," she replied. "Let's load up the canoe and push on to the next island."

Another day passed, but by evening, Anez could not paddle any more.

"My God, you're burning up with fever, "said Steve as he felt her forehead.

"I feel bad, really bad," came the feeble reply.

"I feel woozy myself. We'll pack up and paddle out tomorrow morning." Sleep did not come though, as a horrible ache invaded his back and joints.

Dawn broke. Steve looked at Anez in the dim light, and he could see the broken blood vessels under her skin. A chill came over him as he recognized the symptoms of dengue fever.

In their haste to load the canoe, some supplies were left behind. Anez tried to paddle, but began to vomit uncontrollably. Finally succumbing to nausea, she lay down in the bottom of the boat and fell asleep. Summoning all his strength, Steve paddled through the day and into the night, knowing that their survival depended on getting medical care as soon as possible. Unfortunately, insect repellant had been left at the campsite and by midnight the deadly insects had descended upon him, covering his body with bites, and injecting him with more of the virus.

Morning was accompanied by the sound of warblers, and the noise of an outboard motor. Rounding a curve in the channel, a ranger saw the drifting canoe, its occupants slumped over in a coma. "Good Lord," he exclaimed as he saw their horribly discolored skin and blood coming from Steve's nose and mouth. He had seen this before and knew what it meant.

An ambulance was waiting at the dock but despite medical care, Anez did not wake up. "It's too late," said a doctor. A few hours later, Steve joined his wife in the morgue.

"There are thousands of new cases in the southern states," said George. "Unfortunately, there is no cure, and no vaccine for that virus. Malaria is starting to spread, too, and with mosquitoes breeding in all the rain puddles, clogged gutters and discarded plastic water bottles, it's just going to get worse. I heard that there are hundreds cases of Cholera in Texas. This warmer weather and more rainstorms are just what those insects need. They're spreading like wildfire."

A modest funeral was held at the little Lutheran church where the traditional flowers, organ music, and soothing words of Father Reidesell calmed Stormy, George, and the children. Later, amidst all the food, relatives, and neighbors, Stormy realized that the very climate she had been fighting all her life had taken her mother and father. Instinctively she clutched Jack and Susana to her side.

"I'm going to protect you from this monster," she told them.

And Now the News

> *SUBWAY FLOODS AGAIN, AND AGAIN: This is the third time that the New York subway system has flooded," Stormy wrote, "but efforts to restore it seem to be failing."*
>
> *"Overwhelmed by debt and the second national mortgage crisis, two recent bond measures have failed. And there is no help coming from Washington, preoccupied with a looming oil war in the Arctic and unwilling to provide money to a*

democratic stronghold. In Albany, NY, stubborn politicians would not raise taxes to help the beleaguered city."

Stormy compared them to politicians who had brought California government to its knees twenty years ago.

Polltakers cited the rise in sea level, several hurricanes, and frequent six-inch northeaster rainstorms as reasons for failure to revive the aging structure. It was not just the subway that needed fixing. A big part of the problem was New York's combined sewer/storm drain system, something that was almost impossible to expand due to costs and utilities buried beneath the streets. Designed to carry both sewage and rain water, it was easily overwhelmed by the increased intensity of rainfall generated from climate change. When full, water just backed up and flowed down the streets to the lowest place it could find. Near the subway, those places were the underground tunnels that once carried trains.

"Why spend millions to bail it out, it'll just flood again," was heard over and over again. "Besides, we need that money for sea walls, LaGuardia, and Kennedy airports, the Holland, Lincoln, and Battery Tunnels; they all need protection. The subway isn't just the only thing flooding," said the man on the street. "And this sea level rise is going to continue, there's no doubt about that." Near bankrupt, the City was powerless to act.

City administrators established commissions, paid consultants for plans, and filled shelves with reports, but the flooding continued. A few small and local storm water retention areas had been established, but hybrid taxis still forded flooded streets. Sea walls were proposed, but rejected as being too expensive.

"What is happening with the climate is beyond the boundaries of anything we have experienced," one city planner said in the lobby while sipping his coffee.

"Welcome to the world of climate change," was Dr. Henley's reply.

A private fund has been established and several million dollars were raised for the most recent repairs continued the WGMedia report. Opponents sued, blocking the renovation in court and citing the illegality of using private funds on city property, union concerns, and other "legal" issues. In the end, newly appointed judges sided with the City, throwing out tea party claims and allowing private funds to fix the soggy subway. Once again subway trains traveled beneath the streets of New York City, beneath sea level and beneath the city- wide drainage system designed for a different climate. Nervous passengers checked weather reports whenever it rained.

Stormy Writes - It's Not the Money

In one of her best editorials, Stormy wrote about the root cause of the world's problems.

> *"If you can afford it, you can do it. There is no limit on how much carbon you can put into the atmosphere because you can use the air as a sewer. Decisions are based on how much it costs YOU, not on what it costs the planet, or how much carbon it feeds TERRA. The wealthy, they can still afford eight cylinder engines, 68-degree houses in the summer and all the coal fired electricity they can use. The poor in third world countries and America's slums, well they can struggle to get by, join the radical climate posse, or the climate refugees that are now massing in Dallas, Miami, and Los Angeles. Governments still tax income, but there is no price on carbon pollution; you can buy as much as you can afford. The greatest good for the greatest number: That should be the measure of how we live, not money."*

In her heart however, she knew that before such changes would happen, it would have to get worse; a lot worse.

And so it did.

Uncle Pierre and the Coal Plants

Disappeared, gone: Pierre la France was nowhere to be found. His closet was empty and his even his French flag was missing. Until that is, he showed up on the evening news. Mainstream television did not often cover climate issues, deeming them too trivial for their customers and too outlandish for the companies that paid for the advertisements, but they did cover explosions at coal plants, and when the boilers at three of them mysteriously blew up, it was front-page news. No one knew why, but they were sure it was sabotage. It seemed that bombs were appearing in the coal trains, disguised as lumps of coal.

"They're killing the planet," yelled an unshaven Pierre into the television microphone at a demonstration against coal power.

"Are you responsible?" asked the newsman.

"Everybody is responsible!" he yelled back as a large mass of demonstrators chanted behind him.

In France, the government was not amused that one of its own was embroiled in an American political demonstration, but also noted that France agreed with him that coal plants were the biggest single source of carbon that had been changing the climate. "Mr. La France may be overstating his case, but he is right," replied the French foreign minister.

Undaunted, Wyoming coal companies opened new mines and laid plans for new power plants, flooding the media with lies about "clean coal."

TERRA - In Response to Feedback Loops

TERRA had a busy decade. In ten short years, she ignited massive fires in the drying Amazon rain forest, destroying 80 percent of what had been one of the planet's major carbon storage areas. She claimed her first city as Perth, Australia was abandoned when water supplies dried up and operating desalinization factories became too expensive to supply enough water. Fueled by a lingering El Nino, the Aussie multi-decade drought continued. Undaunted, TERRA raked Pennsylvania with fierce rainstorms, twenty-five inches falling in 48-hours and causing previously unimaginable flooding while destroying dozens of bridges along with thousands of acres of farmland.

Fueled by feedback loops, TERRA seemed to be drunk and out of control. Gone in the summer was the Arctic ice cap that had reflected her heat. In its place was an increasingly acid ocean, unable to absorb as much of the ever-increasing human emissions of carbon. Gone were the rain forests that had stored massive quantities of carbon, giving it up to the atmosphere. Gone were vast areas of frozen tundra that had kept the dangerous methane gas trapped for tens of centuries.

The planet had tipped into a new state; a new system of weather patterns, ocean currents, and storms. And although her average temperature was only 59.3 degrees, increases of several more degrees were now built into this new system. It was a point of no return.

Stormy dreams –A Warning

Tossing in her bed, the dreams came again. Unreal dreams with bright flashes of light, raging fire, and violent water. A white prophet emerged out of the clouds, speaking to her. His words were loud, but garbled, and he seems to be saying "stop, stop." Behind the image, large, black clouds began to gather, growing in size and intensity. More and more stars filled the sky, but they were dimmer, and the ocean had streaks of purple.

Again the voice came. "Stormy. Listen. I am not your slave. Look." Images of forest fires, floods, and famine filled her mind. "You must stop the carbon emissions. My power is great. There will be hurricanes in Los Angeles and in New York. New Orleans, Phoenix, Las Vegas, and many cities will be abandoned. A great famine will envelope China and India. A terrible war will be fought the Arctic."

"But politicians are denying you exist," she replied. "How can we stop the carbon emissions when they don't believe you?"

Sweating, she awoke, quickly dressed, and went to work.

CHAPTER SIX – GLOBAL DISRUPTION
2040 - 2050

Atmospheric Carbon Level - 460-485 ppm

Average Earth Temperature -59.5°
Sea Level Rise - + 2.5 feet
Earth's Population - 8.3 billion people

<u>Stormy's Life</u>
The children were growing up in a climate-changed world. To them, water rationing was considered normal and shortages of food were not unusual as harvests of corn, wheat, soybeans cassava, and rice steadily diminished. Intense rainstorms in the northeastern states were as common as drought in the American southwest, as well as in southern China and India. To them, this was the way of the world, not a distorted, disaster laden, and unpredictable climate that Stormy and George considered it to be. For most people in the southern hemisphere, beans and dry tortillas were considered the standard fare, not steak, and avocados. As oil supplies declined, walking or riding bicycles was a more common method of transportation, not thousand-mile vacations in private automobiles.

As part of the privileged elite, Stormy and George did not want for life's essentials. But they were aware of the increasing number of people who went to bed hungry every night, awoke to parched breakfasts and the sound of gunfire. In third world lands and growing pockets of American cities, people, like TERRA, were becoming restless.

But George, ever the optimist, had plans to do something about it.

"I filed papers today to run for the presidency," he announced. "And you know what, we've got a good shot at it. The polls are in our favor."

As Stormy and George celebrated his potential candidacy with a toast, life went on: including an increase in carbon emissions from a still growing sea of humanity. Jack and Susana joined more than eight million other souls on a planet that was not getting bigger.

"How will it end?" wondered Stormy. "What will their lives be like?"

Christmas, 2040

Twas the summer before Christmas, but no elf could be found.

All across the Arctic, there was no solid ground.

The ice had all melted, it had all gone away.
The North Pole was gone, to the bottom they say.

For the climate had changed, and would continue to warm.
Until even in winter, no ice it could form.

The stockings were hung, by the chimney with care.
But there was no hope, that Santa would be there.

The lights of the cities, were all very bright.
But poor Rudolph's nose, was nowhere in sight.

The air was all sooty, the children were sad.
The Sierra Club was angry, because the coal plants were bad.

I was not asleep, for this is no dream.
The climate will be worse, than it ever could seem.

But I heard Santa say, as he swam out of sight.
"A lump of coal for you, will be mankind's plight."

So email your congressman, ask where will it end?
Slow down the emissions, this is more than a trend.

Sending Corn to Africa

"Susana, why are you putting corn in that box?" asked Stormy.

"I'm sending it to Africa," said her little daughter.

"Why?"

"Because they can't grow enough food there anymore. The rain doesn't fall enough to make it grow. People are hungry, but why doesn't our government stop the bad climate change that keeps them from growing food?"

"It's because of big businesses that have money, lots of money, and stopping climate change would mean they have less money. Our governments cannot deal with something so complex, so they mainly do what the business lobbyists pay them to do, and lots of them don't care about children starving in Africa."

"What about the other countries?" asked Susana.

"Well, the poor countries know they're not at fault, but they are too poor to do anything about it, so they blame us. And you know what? It is our fault. America and China have poured countless billions of tons of carbon into the air, but it doesn't go away. You could say that we didn't know we were creating a problem, but that hasn't been true for half a century. So let's go ahead and send the corn; at least it will keep somebody's belly full for a day. When you grow up, maybe you'll be able to do something to fix the climate so they can grow their own again."

"I will mommy, I will."

A Growing World Food Crisis

Stormy frowned as she scanned her emails and blogs. Reports from around the world reflected a surge in food prices, scarcities in urban areas and growing unrest amidst hungry people.

"Boothby, our readers need to know," she told her boss.

"Know what," he replied.

"They need to know what's happening in China, Africa, Indonesia, Egypt, Mexico, and India. People are starving over there."

Boothby leaned back in the chair behind his desk and looked at her. "Uh oh, I sense a world tour coming up."

"Good, you agree," she said with a smile. "I can leave tomorrow."

"Now wait a minute," he blurted as she ducked out the door.

On the plane Stormy reviewed the United Nations Food Price Index, which had risen to over 400. "It was less than 100 when I was born," she murmured to herself. "Why?" she asked out loud.

A study of grain yields provided the answer and she quickly dashed off the first of her reports.

> *Corn and soybeans produce as much as 10–20 percent less yield for each two degree increase in temperature during the growing season. Planet Earth has already warmed more than two degrees, and will warm even more no matter what we do today because of all the carbon in the air. Increases of 11 degrees are forecast for this century. That will effectively wipe out most agriculture.*

"What is the food situation in China?" she asked the interpreter.

"It is not good" came the translation. "Our water supplies are not sufficient to increase irrigation for rice and increases in fertilizer have not resulted in an increase in wheat production. As a result, we are counting more and more on food grown in African lands that we have purchased."

"China has bought farmland in Africa?" inquired Stormy.

"Oh yes, and in South America too. We could not feed our people without doing that. That dust you breathed in Beijing used to be our fertile soil, but overgrazing, drought and soil erosion ruined it."

"The air was so bad it was hazardous. I had to wear a mask outside," replied Stormy. "How often do you have those storms?"

"Every year now," said the lady with a sad face.

The Chinese government answers were duplicated in Africa where Stormy asked the same questions to workers in Ethiopia and Sudan. Stormy wondered why these countries, which were receiving food donations from the United Nations, were selling and leasing their agricultural land, but responses from agency officials were indirect or totally lacking. It seemed that money was involved.

In Nigeria, she asked: "I know your livestock production has increased, but aren't you worried about how that is affecting your grasslands? It seems the land is overgrazed, the desert is spreading and only goats can live there."

"Desertification is our number one problem" said her guide. Dressed in a turban and white robe, he pointed to the vast area of shrub land before them. "This used to be grazing land for cattle. Now the soil has blown away and only shrubs can survive."

"There were too many cattle, and too many sheep," said Stormy.

"I know that now, but we have too many people to feed, and hungry people are hard on the land."

Filing her report, Stormy wrote:

> *Several African countries are quickly turning into deserts. As the population grows, the ability of the land to support them is shrinking, and climate change is making it worse. These are not nations with a future.*

In India, the situation was similar, but worse.

"More than a third of our nation has become a desert," explained the Indian scientist while showing her images from space. Ground water levels are declining and our sacred cattle are skin and bones."

It's not just the cattle that are starving. Stunted children can be seen in every slum where foodless days are common for many families. Can you imagine going for a whole day without food? And with the continued decline in glacial water used for irrigation, a full scale famine is imminent.

So wrote Stormy from the hotel room in Calcutta. Her travels were interrupted however, by a more immediate problem. When she returned home, there were other issues that demanded attention.

World at War

"Stormy, your article on flooding in the New York Subway was nominated for a Pulitzer Prize," said George, as he walked in the door.

"Yeah, but the winning feature is going to be the report on skirmishes in the Arctic over oil," replied Stormy. "Look at this."

She threw the evening headlines on the table. George picked up the paper and read:

US / CANADA COALITION DECLARES WAR ON USSR OVER ARCTIC OIL!!!
Dateline Moscow: The Soviet Union/China coalition responded to the declaration of war today by placing military bases on full alert and mobilizing troops. Elsewhere, third world nations in Latin American, Indonesia, and Africa joined the Soviet/Sino coalition. Condemning the U.S. for the blatant attempt to gain control over arctic oil and for creating the greenhouse gases that have destabilized the climate, the impoverished nations of the world are turning against the United States.

"I was expecting this," he replied. "Billions of dollars are at stake and the lobbying is fierce in Washington. I've never seen anything like it. The United States still refuses to sign the Law of the Sea treaty, like every other country has done. Sarpin and her allies just won't back down; they want that oil. The military thinks they can win of course and people on the street think it will mean lower fuel prices. We may be in for a long war."

"The real war is worldwide, with TERRA." replied Stormy. "This oil stuff is just one of the isolated battles with a climate that is getting worse, everywhere. I just hope Jack won't get caught up in it when he grows up. Are they going to start drafting for the army again?"

"No, this war will be fought with guided missiles. Oh, the solar panels were installed today. We are officially off the electric grid," beamed a proud George. "At least we are doing our part."

"What individuals do isn't enough," said Stormy. "Governments need to put a price on carbon so market forces can shift us over to renewable energy instead of fossil fuels. Instead they're going to war so we can burn even more carbon. Come here, I need a hug."

Conversations in a Pub

Boothby lingered at the bar, sipping his Irish coffee. George and Stormy had joined him for his birthday celebration. "It's the damn government's fault!" Boothby said as he burped, swinging his drink high in the air.

"Right." replied Doth the bartender. "I don't like those politicians."

"Nor I," bellowed Boothby. "They can't get anything done. All they do is argue."

"We agree. I'd fix it if I were in charge," said Doth, drying a glass: "Especially if I had all that tax money."

George smiled and sat back, waiting for the other shoe to drop.

"The money's being used for the wrong things," said the Booth while paying for his drink. "We need to use it to stop climate change, not buy more oil."

"Stopping climate change would mean going back to the Stone Age," said Doth, "and that would be stupid. We need progress, growth."

Staring at each other, they suddenly realized that they both hated politicians, but were on opposite sides of the tax argument.

"Right," said Boothby, sarcastically. "But what was wrong with the life your grandparents lived in the 1950s? Were they impoverished because they had only one car? Did they suffer because their food was grown locally and not shipped in from 1,000 miles away? Were their lives poor because they didn't have air conditioning, two bathrooms, and four bedrooms?"

"Well, I suppose not."

"We don't need to go backward," said Stormy. "We just need to replace our energy supplies with technology that doesn't pollute *and* we need to limit our population growth."

"It doesn't matter," said Doth. "God said no more floods, fire next time, so there won't be any climate change floods."

"But what if God is wrong?" said Stormy.

Doth stood there with his mouth open for a moment, sputtered and, not finding any words to answer her, went back to drying glasses behind the bar.

"You know, it's like a worldwide poker game," continued Boothby. "Humans hate to lose, so every time nature puts up a barrier, we raise the bet. Too many people think we can just conquer TERRA and keep burning fossil fuels. But in the end, the laws of physics can't be changed. And, sadly, the people who know what is happening are not the wealthy ones that control politicians. I'm too old to see most of it, but I feel sorry for our kids."

"Science has always been ahead of its time," said Stormy. "Galileo was damned for believing the earth revolved around the sun, not the other way around. When Columbus sailed off into the horizon, many people thought the earth was flat. It took hundreds of years for the average person to believe what science told them rather than what their senses or religion told them."

"Unfortunately it's also that way with climate change," said George.

"This has happened before," said Dr. Henley who had just joined them. "But it was a long time ago."

"How long ago?" said Stormy.

"Way back when," he replied. "You know, before the trans-ocean invasion. Around the time that rocks were invented and Moses created horses."

"You're drunk Dr. Henley," exclaimed Stormy. "Come on, I'm writing an article on this. Millions of people are going to read it and it's got to be right."

"Just say that it was before people were on the planet," he said. "Civilization has never seen anything like the climate we are creating now. Dinosaurs maybe, but not people. Did you know that, sixty five million years ago, the temperature of the north polar sea was 73 degrees?"

"Seventy three degrees!" exclaimed Stormy: "In the Arctic Ocean?"

"Correct. Our climate system did it before and is still capable of that now. If it is that warm up there, just imagine trying to grow crops in Kansas. You can't: No wheat in Kansas. My God, we'll starve."

"We've simply got to reduce carbon emissions," said Stormy. "What will happen next if we don't?"

"This climate thing is like Vietnam," said Boothby. "People need to really see it, and see how bad it is before they are willing to do something about it," he went on. "Hey bartender! Have you ever seen those bands of climate refugees carrying water in jugs on their heads?" he asked.

"Nah, with all the bad news out there I just read my sports blogs," he replied. "I don't believe in that global warming stuff anyway. Those scientists don't know what they're talking about. And that Nobel Prize they gave away isn't worth crap. I found one in my box of Cracker Jack last week."

Stormy grew furious.

"Go lemmings go!" she yelled. "Keep on going until you go off a cliff. What about the melting ice caps, billions of dead trees, the heat waves that are killing thousands of people, the rising sea level, and bigger hurricanes?" She was really mad now. "You god damned deniers; your opinions are based on politics, not science. Your words are based on greed, insensitivity, and selfishness."

"Calm down," said George.

"What about the wars in Africa and India because monsoon rains have shifted course; and people on islands that are being flooded out of their homes? Why are the oceans becoming more acid? So tell me, if the scientists are lying, how do they coordinate all those things at once?"

Stormy's ranting had drawn attention from a new group that had just entered the pub. It was the Climate Posse, a band of radicals dedicated to destroying the technology that was creating the climate crisis.

"I've heard of you," said Boothby. "You were in Wyoming last week; blew up a coal train if I remember right."

"Shut up old man."

The bartender threw up his hands and turned away, but Stormy's yelling started a fight at the other end of the bar as the deniers and Climate Posse started swinging. A chair flew across the room and hit the jukebox.

"Dr. Henley will tell 'em; he knows that climate change is real. Tell 'em Dr. Henley!" screamed Stormy as a half full glass of dark beer was thrown toward her and a patron threatened her with a knife.

Dr. Henley ducked out the back door.

"Sean, get your shillelagh!" yelled Boothby as the owner grabbed a pool cue and threatened two customers. But it was too late. Passions were high as musicians joined the fray and students took up the challenge. The bartender landed a punch, but was wrestled to the floor.

A denier threatened Boothby, pulling her arm back for a punch. Stepping up beside him, Stormy grabbed the bottom of her hardback book with both hands, whopping the hapless drunk on the side of her head, knocking her to the floor and causing her arms to rise in supplication.

"Wow, that book packs quite a punch," said Boothby. "What is it?"

"Al Gore's latest book," replied Stormy. "*An Assault on Deniers*."

"Look out!" cried Boothby as someone raised a bottle of Jamison's over Stormy's head.

Fortunately, Sean was there to grab it. "I'll not waste good Irish whiskey," he said: "beer maybe, but not Irish whiskey. This climate crap is bad enough without you all fighting about it," he said. "Go back to your drinks."

Things settled down when Sean turned on the sprinkler system and played classical music at full volume. Un-noticed, uncle Pierre ducked out the front door. No one ever heard from him again, probably because the U.S. Secret Service had issued an all-points bulletin regarding his capture, reading: *Shoot to kill.*

"At least it's not as bad as in Africa and China," said Stormy who had calmed down. "Civil wars are raging over there. It doesn't seem fair. Here we are in the middle of a paradise with fine wines, imported food from around the world, an abundance of wildlife, and lots of leisure time while billions of people just like us are starving."

"That may well our fate," George replied. "In a few decades, climate change will impact the lives of everyone, including the wealthy. You're too young to remember the dust bowl, but it affected almost every American from 1930 to 1936. That drought ended, but climate change can produce permanent drought. There are rusted Massey Ferguson tractors abandoned in the fields right now. There are half empty silos in Iowa. The Texas climate is like that of southern Mexico and even the grain belts in Russia and China are drying up. What will it be like after twenty more years of drought?"

George sat in his favorite chair, thought a while and said, "when I get to be president, I'll certainly tackle it. The campaign is going well; we picked up key endorsements from the unions, teachers, and a few Hollywood stars. But fund raising is still a problem."

"Enough pessimism," said his wife. "I propose a toast: May everyone be always as happy as we are now."

"To happiness, I hope."

Congressman George

Back in Washington, George was busy writing the "Emergency Climate Act" that he introduced on Earth Day, April 22, 2045. Known as the holy grail of response to climate change, it would phase out all coal burning power plants, build safe, fourth generation nuclear plants, and tax carbon at the source. Billions of dollars would also be allocated for wind, solar, and geothermal energy.

"Bullshit!" cried supporters of Sarpin and the climate deniers. "We've got to stop this."

At a late night meeting in a bar near the beltway, they decided it was time to take matters into their own hands.

"We need to send a message this George fellow," muttered the former mid-western senator.

"And I know just the guys to do it," came the response. "Just make sure to keep me out of it."

With an exchange of money, the plan was set.

"Congressman, may I have a word with you?" said a shadowy figure on the capitol steps.

It was near midnight, but George stopped to answer, wondering what a news reporter was doing out at this late hour.

"It's him all right. Bring the car around."

"Who are you?" asked George.

"Your worst enemy: Get in!" he yelled as he held a short barreled 38-caliber pistol to George's stomach.

Blindfolded, George tried to keep track of the turns and traffic patterns as the SUV sped by dark capitol buildings, finally stopping below a bridge on the Potomac River.

"Get on the boat," said the burly man with a gun. "Or taste the bottom of the river."

After a short ride across the water, George found himself in a shack, tied to a chair and facing a bright light. He could not see his protagonists.

"You are not going to pass this climate crap act," was the next demand, apparently spoken by a young woman who seemed strangely out of place. She left the room, leaving George alone to wonder about his fate, and how he could get out of this. Outside, he could hear arguments.

Morning dawned and the young woman appeared again, this time carrying food, water and a manifesto that she demanded George sign.

"I can't sign this," he said. "It's not true, and besides, nobody would believe it."

"Why not?"

"This stuff about how global warming is a natural cycle is bunk," said George. "That was disproved decades ago. What's happening now is because people are turning TERRA into a runaway train."

"TERRA, who's TERRA?"

"The climate. I know it sounds theatrical, but people respond to something they can name. Haven't you read Stormy's blogs and editorials?"

"Oh, that bitch," came the reply, causing George to turn red as he struggled to contain his anger over hearing his wife called a bitch.

"Look, even the weatherman on channel 14 admits that solar radiation alone can only account for a fraction of the heat that satellites have been measuring."

The satellite argument seemed to get her attention, and George continued this line of reasoning, hoping to wear her down. As the day dragged on, the Capitol police were at work.

"He's been WHAT?" yelled Stormy when the call came in from Sergeant Blesy.

"Kidnapped. I saw it on a security camera video. When we got the emergency signal from his phone, we tried to stop them, but it was too late. Our patrol car lost them on the freeway."

Three days passed, but George would not sign the denier's manifesto.

"What are we going to do with this guy?" muttered a bearded man outside the shack where George was being held.

The young woman paced up and down. "I don't know. What does the boss say?"

"He skipped town when the heat came down and nobody knows where the bastard went. The fuzz bugged his phone and they're on to us. But I'm not taking the rap for it."

"Me neither," said the woman. "And besides, I didn't know he would be a nice guy. He really does believe this climate stuff."

"Yeah, you're getting soft."

"Let me try one more thing," she said, moving slowly into dim room where George remained tied to the chair. "Hi George," she smiled.

Her sudden friendliness startled him. "Uh, hi," he managed to say.

"Look, you're a nice guy, good looking, too. Why don't we work something out?"

As she spoke her hands moved to her blouse and began unbuttoning the tightly stretched fabric.

"Oh, no," thought George, realizing what was coming. Insulting her would be a mistake. He couldn't say that she was nowhere near as beautiful as Stormy, or that her approach to sex was candidly artificial. So he said "wow" as the bra fell to the floor, followed by something that made Stormy laugh when he told her later. "Don't tell anybody in the Senate, but umm, I'm gay."

With a shocked look on her face, she pulled her blouse back on and stormed out. "Okay, cut him loose. We're out of here."

Reunited with Stormy, George became a celebrity and became the nation's candidate to become president of the United States.

Fame however, did little to help his climate legislation. Millions of dollars from oil and coal companies were spent to defeat it and discredit the science behind it.

"I wonder if the people putting up the money were the same ones who kidnapped me," said George as he launched an investigation. Sarpin's political party countered by claiming that George had faked the entire kidnapping, which made Stormy furious. George had to remind her that it was just business as usual in Washington. But as the decade ended and the climate emergency act was forgotten, headlines continued to pour in from around the world.

And Now the News: Dateline 2048

> *CLIMATE IS NOW OUT OF OUR CONTROL: World scientists claim that climate change is now past the tipping point and beyond our control, being driven by feedback loops and an insatiable appetite for fossil fuel. Economists state that there is not enough money in the world to build barriers for*

sea level rise, much less to pay for starvation, disease, and storms.

U.S. FOOD PRODUCTION AT AN ALL TIME LOW: As the multi-decade drought increases its grip, farmers in California have proposed pumping more ground water for their crops, but with reduced snow melt, less runoff, and limited hydro power, electricity is not available to run the pumps. In the mid-west, corn and wheat yields have sunk to dust-bowl era levels and thousands of farms have been claimed by banks. In the southwest, native people prayed to Kokopelli, who played his flute over the parched land, but the rains never came.

MILLIONS DIE OF STARVATION: Crops failed again in Africa this year and food rationing has begun in China. The Fertile Crescent, once the breadbasket of the world, has dried up, but imports are not available. World protein consumption is at an all-time low and prices of food have reached all-time highs.

DENIERS RESPOND: "A few million people died: So what?" The paper was quoting Cedric of the People's Association for a Sane Earth. "You enviros have been saying there are too many people on the planet already. Losing a few million here and there doesn't mean that climate change will destroy the Earth. My air conditioner still works and I can still buy gasoline. Climate change is somewhere else."

SEA WATER FLOODS THE NILE DELTA: Land subsidence, sea level rise, and reduced flows from the Aswan Dam have caused a surge in salt water in the lower delta, threatening agriculture, tourism, and ground water supplies. Worldwide, sea levels are now almost three feet higher than in the last century.

HURRICANES HIT RIO, TOKYO, CALCUTTA, AND LOS ANGELES: Expanding their range and intensity, hurricanes have struck cities never before hit by such monstrous storms. Climatologists blame rising seawater temperatures due to greenhouse gases. Tens of thousands of people have died and millions have been impoverished as losses to homes, businesses, and infrastructure continue to mount.

LAKE MEAD DRIES UP: Flows in Lake Mead were reduced to a trickle in October, causing the city of Phoenix to issue water rationing cards for residents, limiting them to 10 gallons per day.

LAS VEGAS REAL ESTATE MARKETS COLLAPSE: Continuing brown outs, heat, and lack of water have caused thousands of people to flee Las Vegas. Property values have plummeted to unbelievable lows in a city that is rapidly shriveling into a ghost town.

DENGUE FEVER AND MALARIA SPREAD: Epidemics of climate related diseases continue to spread as cases are reported in Virginia, Maryland, and New York. A shortage of medical supplies has hampered control efforts in central New York City.

HUGE STORM RAKES MIDWEST: It began over southern Missouri. Fueled by moisture from the Gulf of Mexico, cooled by arctic air from the north, and intensified by a low-pressure system, the storm of the century dumped eight inches of rain on Joplin, Missouri, site of a previous deadly tornado. As the cloudbursts gathered strength over Illinois, they hit Chicago with a tremendous downpour of three inches of rain an hour that lasted for three hours. Damages to utilities, residences, and infrastructure were estimated to be in the billions of dollars. Hundreds of tornadoes intensified the destruction.

NATIONAL PARKS IN ASHES: As had been predicted for years, lightning storms ignited forests in Rocky Mountain, Glacier, and Yellowstone National Parks where old growth timber had succumbed to bark beetles. The tiny insects, now surviving the warmer winters of a changed climate, have devastated vast areas. The national park, where John Denver once sang Rocky Mountain High, is now in ashes. Only a blackened, barren landscape exists where once campers lounged in view of majestic peaks. Soil eroded from unprotected forest lands now covers roads and most park buildings have burned to the ground. "At least it killed the damn beetles!" exclaimed a frustrated tourist as he toured the devastated landscape.

Jack Moves to Canada

Precocious like his mother, Jack left home at an early age, joining the military which sent him to the Canadian border. There he saw the hundreds of thousands of Americans and other wealthy world citizens moving to southern Canada.

"It's really great up here," he told Stormy on the satellite phone. "Edmonton, Calgary, Saskatoon, Winnipeg, Vancouver, Toronto and Quebec are really booming. I know the oil boom is bad for the planet, but people are really raking in the money from it and folks from the states and all over the world are flocking to the cooler climate."

"How far north is this boom extending?" asked Stormy.

"Not far," he replied. "There's hardly any infrastructure in the northern mountains, plains or the rocky wetlands of Ontario, but the cities within 100 miles of the border have so many newcomers that they can't handle them. Salaries are sky high, housing is hard to find, utilities are overloaded and the roads are crowded. "

"Go further north son," said George, "to the Arctic."

"It's booming up there too, in the summer anyway," said Jack. Tourism, mining, and oil/gas development have all been spreading along the Arctic coast since the ocean became ice free for several months. But there REALLY isn't any infrastructure to support all this. The first cruise ship dock and customs station for Barrow, Alaska was just built only a few years ago. Before that you could dock a ship or fly in without anyone knowing about it. It's the new frontier."

Stormy could sense the excitement and enthusiasm in his voice but her motherly instincts were also aroused. "Be careful up there in the Arctic," she said. "There are lots of ways to get into trouble without any way to get out. And be sure you don't do anything to make it worse for the native people still trying to live off the land. They deserve our respect."

"Agreed," said her son. "I'm planning a winter trip offshore to the remaining ice fields to see if there are any polar bears or narwhal left, but I'll be careful."

"Send us a video," said George. "See you soon."

TERRA - World Wide Anger

Now thoroughly unstable, TERRA reached into every part of the planet, creating more extreme weather everywhere. As she spread third world conditions across the continents, economies, and governments destabilized, and threatened to collapse. Algeria, Syria, and Mali were already ruled by anarchists. Greece, Spain, and North Korea joined other nations in bankruptcy as international loans dried up and resources were diverted to fight sea level rise and diminishing food supplies.

In the tropics, TERRA destroyed ninety percent of the formerly magnificent coral reefs. Bleached by the warming seas, their broken skeletons littered what once had once been thriving reefs with millions of fish. In their place, divers reported massive colonies of jellyfish, floating mats of algae, scum, and bacteria. Tourism to these areas plummeted.

Spurred on by hundreds of thunderstorms, TERRA threw lightning strikes everywhere, each degree of temperature rise resulting in millions more of the potentially deadly flashes. As seen on satellite photographs of the central plains and in Canada, their frequency increased dramatically, as did deaths from the deadly strikes.

Increasingly, TERRA was drawn to her past, when climates had been extreme and alligator like reptiles swam in arctic latitudes. In the millennia since then, ice ages had come and gone, continents had drifted, ocean currents had shifted, and countless living plants had formed fossil fuels that were buried in the Earth. But now these centuries of carbon deposits were being burned in mere decades. Nothing like this has ever occurred since the planet had been created. TERRA was angry.

Stormy Dreams - Anger

Stormy tossed and turned in her bed and as her dreams became more intense, the world began to spin, and then to shudder. A little girl in a white dress seemed to be asking for something, but what? Flashes of light sparked through the black clouds and the ocean grew violent and black. The Earth was cluttered with people and massive numbers of stars glowed dimly in a reddish night sky as flames enveloped the brown forests. Mountains receded in the background, fading into oblivion.

As TERRA became even angrier, Stormy's dreams grew increasingly violent. She wanted to lash out at the forces that were crippling cities, homes, and people as if they did not matter. But, being unable to fight a force that she could not see, Stormy directed her anger at those who were responsible for feeding climate change. Unchecked by a waking world, she drove her car into groups of climate deniers, spit upon congressman that continued to subsidize coal companies, and kicked the idiots who drove their huge vehicles into three car garages. Stepping on their hands clutching keys to oversize cars and six bedroom houses, she screamed at them and threw their right wing propaganda into a fire.

"Damn you, damn you to hell!" she shouted at the television announcer interviewing a faux scientist wearing a laboratory coat who was implying that solar cycles and mysterious chemicals in the atmosphere were responsible for the storms. "Don't you ever read?" she asked the big screen. "Can't you think for yourself?" asked the angry young woman.

Cursing, she threw a shoe at the flickering box, silencing it as heavy rain pelted her windows.

"Fight them Stormy," came the voice, deep and hidden in the sky. It was as if TERRA was speaking to her.

"You are the hope of the world. Fight for the Earth."

CHAPTER SEVEN – METHANE TAKES OVER
2050 - 2060

Atmospheric Carbon Level - 485-510

Average Earth Temperature -61 $^{\circ}$
Sea Level - + 4 feet
Earth's Population - 8.0 billion people

Jack Dies

Come all you young rebels, and list while I sing,
For the love of one's country, is a terrible thing,
It banishes fear, with the speed of a flame,
And it makes all part of, the patriot game.
"The Patriot Game" by Dominic Behan

"We're going down," shouted the pilot. "I can't hold her."

Spinning like a top, the galaxy class helicopter wobbled, and then started to plunge. Ice had formed on the tail rotor and there was no way to clear it. "Over there," shouted the co-pilot, "on that iceberg." As he jammed the stick hard to the right, the big machine slowly responded, still spinning but now moving toward the ice flow and not the black arctic sea beneath them. "We won't make it," was heard from a crew member behind them while the water loomed only ten feet below. But then, as by fate, a strong gust of wind hit the craft. It lurched, and then crashed hard onto the ice.

"Thank God we're alive," thought Jack on his first military mission to the Arctic. Strapped in the rear of the chopper, he stripped off the safety harness, but being dizzy from the ship's movement and overwhelmed by panic, he sat still for a moment. It was a deadly mistake. Thinned by decades of melting, the once thick ice was not strong enough to hold the aircraft. Giant fissures opened to the water below, accompanied by loud cracking sounds. Quickly, within seconds, the green helicopter vanished between chunks of broken ice. It was 4,000 feet to the bottom, and there was no escape. A few moments of panic, an icy chill, and his life was over.

In Washington, reporters blamed it on the weather, but Stormy knew that TERRA had claimed another victim, her son. She responded with an article driven by emotion and the scars of a mother who has lost her only boy.

Stormy Writes - One Planet; One People:

"Jack, why did you die? For eighteen years you walked this earth a smiling, strong young man. Now you are at the bottom of an icy sea. And for what: To fight in a foreign land defending the profits of the wealthy: To defend our honor? What honor is there in fighting for oil deposits that are destroying the earth?

Who will win? Will we continue to chase fossil fuel deposits of coal, oil, and gas to the ends of the earth, fighting each other for control of their greenhouse gases and material wealth? Or will we realize that solar, wind, and geothermal energy are free after the infrastructure is built. What should we do? Send soldiers to defend oil deposits? Or use that money to build windmills, solar panels, and battery powered cars?

If we can pull together, if we can stop the fighting and focus on this crisis that threatens all of us, we might have a chance. But if we continue this senseless war, others will make this same sacrifice, the worst one a mother can make. I miss you Jack, and I promise to work to see that your life was not in vain: Your loving mother, Stormy."

Six months of grieving soothed her soul and, encouraged by George's decision to run for the presidency, Stormy finally returned to work.

A Tightening Grip

A vicious downdraft grabbed Stormy's umbrella, inverted it, and broke the spokes. "Damn it," she muttered. "They can't even make an umbrella that will hold up in these storms." Monstrous raindrops pelted down, soaking through her thin jacket to her skin and making it hard to see across the street.

"That looks like a wall cloud," said her husband, "the kind that spawns tornadoes. We'd better duck in here." George was a practical kind of guy and Stormy had learned to trust his instincts.

Entering the lobby of a high-rise, she remembered the first level six tornado that had twisted the frame of a skyscraper in Kansas. Ten more of these monsters had formed since then and she fervently hoped this would not be another one. "Not here; not today," she prayed. Inside, they were drawn to the wall-size screen now showing local weather radar. An orange blob hovered over the center of the screen, surrounding an inner, red zone and central to that, a big purple mass.

"That's where we are!"

"I'm afraid you're right," replied George. Looking outside at bent over trees and airborne soft drink cans he motioned her away from the large, plate glass window.

"We interrupt this broadcast to issue a tornado warning." The announcer's face looked grim as he shifted his gaze to an incoming call. "Carl here, south of the city," came the voice. "I'm in between two tornado tracks, one a mile to the east and another just off to the west. It looks real bad."

Some hot coffee in the basement of the restaurant eased their fears and twenty minutes later, the storm passed, its damage having been done elsewhere. Summarizing the day's weather in her column, Stormy wrote of the hundreds of tornadoes that had been tracked by satellites.

"Another record broken," she thought. "But I'm getting tired of just writing about the number of storms, the destruction, and dead people; there's more to it than that."

Weeks later, while accessing satellite information from the new Federal Office of Climate, she was astounded at the destruction caused by the most recent mega-storm. More than six thousand homes had been destroyed, sixty million dollars in crops had been lost, and six hundred people were dead. "It was the storm of the sixes," she wrote. As was typical in such storms, several entire towns had been devastated; local utilities were in ruins and for many, insurance was unavailable, having been depleted by the damage from previous catastrophes. Digging deeper into the agency website, Stormy noted how the computer predicted a twenty percent rise in regional poverty, a fifty percent rise in prices of scarce building materials, lower property values, and of course, nationally, another drop in the stock market, which was already at a twenty year low. Within days, climate refugees from the area were massing together and overwhelming local shelters.

> *"World-wide, the situation is unbelievably worse," she wrote. "Although data is sparse, it is now believed that are 500 million climate refugees, one out of every fifteen people on Earth. They have left their homes for various reasons; some to escape wars over water, cattle or land; some seeking jobs after going bankrupt; some fleeing religious persecution and some because their homes were devastated by floods or drought. Most are from dozens of failed nations like Somalia, Pakistan, Iraq, Indonesia, and Mexico between the tropics of Capricorn and Cancer, but increasingly, they are on the move in northern nations. Former soybean farmers from Alabama and wheat growers from Texas crowd into cities while businessmen from El Paso flee drug violence. In Europe they line up seeking jobs. Once displaced, chaos rules and they turn to smuggling, drugs, prostitution; anything to survive."*

"This really is all related," her report continued. "Raging climate disrupts lives; more lives than can be helped by government relief agencies or non-profits. Christian Care, the Red Cross, and Climate Relief have all gone bankrupt trying to help the millions of people whose lives had been damaged by floods, hurricanes, tornadoes, heat waves, and wars induced by famine. Donations, she noted, are all but non-existent as more and more people watch their investments plummet in value and prices rise exorbitantly. But as civilization fades, greenhouse gas emissions and world temperatures continue to climb."

Fleeing Florida

"What's this?" asked Stormy.

"Your ticket to Florida," replied her boss.

"Boothby, you're sending me to Florida? Why?"

"Hurricanes and insurance, that's why. The state is going bankrupt because they agreed to be the insurer of last resort after companies with more sense stopped offering insurance. It's just too risky."

"OK, I'll bite. What does that mean?"

"Well, way back in 2004 - 2005, when you were a little baby, five big hurricanes hit Florida. The damage was so bad that most insurance companies drastically raised rates, or stopped selling policies altogether. They lost billions of dollars, and when they realized it would happen again, they backed out."

"But how can you build anything or even buy a house without insurance?" asked Stormy.

"You can't. And that's the problem. Growth and development was stopped in its tracks, and these are what make the economy grow, or, in this case, not grow. State politicians tried to sweeten the pot with subsidies and get them to write more policies at lower rates, but when the insurance companies ran the numbers, it wasn't worth it."

"What happened then?"

"Well, Florida, being a state with lots of deniers, decided to get into the insurance game and passed laws saying they would provide insurance if people couldn't get it anywhere else. I guess they didn't believe that global warming could really create such risks."

"Uh oh, four hurricanes have already hit this year," said Stormy. A frown slowly filled her face as she realized the implications.

"Right: It started in May with a category 2 in the panhandle, then a 3 hit Miami and another pounded the Tampa Bay area, followed by a category 4 in the Keys. That one killed 800 people, mostly those that chose to ride it out. The drunks were seeing alligators in their kitchens."

"You're sending me into the lion's den! What if another one hits?"

"You're a climate reporter; you report on what climate change does to people. Besides, it's nice and warm down there now."

"You mentioned bankruptcy. Who is going bankrupt?"

"The state government of Florida," replied Boothby.

"Why? Was the damage too great?"

"Oh, it was more than too great. Tampa Bay had 60 billion in damages. Combined with the other three storms, state taxpayers are on the hook for 50 billion in uninsured losses. The entire state budget is only 70 billion."

"Ooops."

"People down there are really pissed at the politicians that got them into this mess and there is talk of removing the governor. Enough: Your plane leaves in two hours. Go Stormy go."

The flight into Miami was rocky as winds from a tropical storm in the Gulf buffeted the airliner, but the landing was smooth. Feeling guilty for riding a machine that emitted such huge quantities of greenhouse gas, Stormy bought carbon credits from a vending machine at the airport, hoping they would pay for preserving an acre of the beleaguered Amazon rain forest and the carbon it had stored away. She checked into a motel with a gorgeous ocean view, chatted with George, and quickly fell asleep.

But as she slept, the tropical depression began to grow. Stalling over hot Gulf waters, the spinning gyre quickly grew into a full scale hurricane, shifted course and headed straight toward Miami and Stormy.

She awoke the next morning to dark skies, strong winds, and warnings on the local news. Residents were warned to seek shelter inland as the storm grew in strength, first to a category 4, then 5 with winds of 160 miles per hour. Storm surges of ten feet were predicted.

"My God, it's here," she said to herself. "I'm in the path of a monster hurricane." Quickly dressing, she grabbed her bags and headed to the lobby where crowds of people had gathered.

"Taxi!" she yelled. But there were none. Rental cars were gone, too, and bus service had been suspended. She went back to the motel to ask a bellhop how to leave. He didn't have any more information than she did.

Outside, the storm began to pummel the motel. Palm fronds filled the air, along with pieces of roofing and anything else not tied down.

Retreating to her room, Stormy considered hiding under the bed, but rejected that idea. That's what a little girl would do, she thought to herself. Pushed by ever increasing winds and rising tides, the ocean waters gathered themselves into a ten foot wave, pulsed across motel landscaping and picked up a picnic table, some chairs, and other debris. Moving faster than a man can run, the debris smashed through Stormy's ocean view window.

Stormy dove under the bed.

As quickly as it hit, the water receded from her room, leaving a dripping morass of motel furniture and bedding. "I'm so out of here!" she screamed.

Drawing heat from hot gulf waters and spinning counter clockwise, the storm moved slowly across central Florida with winds striking the low lying city from the west. Waters from torrential rain and the Everglades began to flood into the city, overpowering the storm drain system and filling streets, sidewalks and low lying homes. As power lines fell and electricity failed, water and sewer plants shut down.

Sheets of unrelenting rain fell while underground in the porous, limestone soil beneath most of Dade County, groundwater rose higher and higher, reaching the surface and adding to the flood.

Stormy walked down to the lobby and huddled with others. Dozens of people sought refuge there in the dim light as murky water swirled around their feet and a deafening roar filled their ears.

"Don't cry," said Stormy to the child sitting next to her.

"I'm not crying," she replied. "The ceiling is leaking."

"Oh my god," said Stormy as she looked up. Water dripped onto her face as the walls shook. The entire structure seemed to be moaning.

Fear turned into panic as the roof, and then the ceiling blew away, drenching the crowded mass of people.

"Hold on to me," cried Stormy to the child as the walls shuddered, buckled and succumbed to the wind, blowing away in pieces. A surge of water enveloped them, carrying struggling people outside into a street flooded with several feet of water.

The child screamed as the wind and flowing water pulled her away from Stormy's grip. Horrified, Stormy could only watch as her face disappeared in the storm. Instinctively she grabbed a floating table and held on as terror enveloped her. Nearby she could see people screaming but could not hear them over the deafening roar of the rushing air.

Past the street the water was deeper and the table began to roll, forcing her underwater, but she could not touch bottom. Gasping and choking she managed to scream "help" into the raging hurricane.

Then, within minutes, the wind ceased while an eerie calm enveloped the area. The eye of the hurricane was passing over her, providing a few moments of quiet as Stormy drifted with the currents, heading toward an open beach where eight foot high breakers pounded the sand and threatened to drown her. A body floated by. She averted her gaze.

Fleeing Florida

Quickly the wind resumed its assault, striking from the east now as the storm moved south past the southern tip of Florida and forced a storm surge from the sea which carried Stormy back into the now flooded city. High rise buildings were on the horizon, but she could not reach them. For hours she drifted past flooded buildings amidst floating wreckage of what had been people's homes only hours before. Darkness enveloped her but there were no lights as the wind and rain continued to pound at the sea of water around her.

Time passed, draining strength she needed to keep her grip on the life saving table. Occasionally something would bump against her beneath the water, causing Stormy to kick it away in fear. Finally, near dawn, her feet touched land and she was able to wade through receding waters with a diminishing wind at her back. Few landmarks, or even anything resembling a recognizable structure were visible. Emerging from the water, she began to find other, shocked people that had somehow survived the storm. By evening she had found safety on the spit of southern Florida that was twelve feet above sea level and had not been flooded as bad.

Looking around she saw once expensive cars on their sides or upside down. Debris and boats filled the street and then she saw a body, lying motionless under a shattered palm tree.

Now famished and dehydrated, Stormy asked "is there any water," to one of the masses of people who had gathered there.

"Look for the helicopters," said a disheveled old man. "They are dropping food and water until the roads can be re-opened."

That day and another night passed before Stormy and other survivors were rescued. It was they said, the biggest hurricane to ever hit Miami, a city situated only a few feet above sea level. Damages were estimated at 50 billion dollars, a large portion of it uninsured. Depleted by previous storms that year, state government coffers were unable to pay any of the destruction as had been promised. The resulting law suits would continue for years after the great State of Florida formally declared bankruptcy. Climate change had claimed another victim, and everyone in the state of Florida was going to pay for it.

Recuperating at home in Washington, Stormy realized that although she had survived, Miami had not. With its infrastructure and utilities crippled beyond repair, she knew that the city would never recover. Sea walls were useless where the rising ocean simply flowed in through porous ground beneath, bubbling up ever higher in low areas and making drainage all but impossible. To make matters worse, a nearby nuclear power plant on the Florida Keyes was decommissioned when it was realized that the combination of sea level rise and stronger hurricanes made it vulnerable. Her views on the situation were confirmed a few months later when another hurricane threatened the city, striking only 60 miles north in Tampa.

"In the future people will be snorkeling among the ruins of abandoned high rise buildings and the Everglades will be a marine reserve for sharks instead of alligators," she told George.

And Now the News: Dateline 2056
Back in her office and focusing on climate related stories, Stormy reviewed events of the week around the world.

> *CROPS NO LONGER SURVIVE: Across vast areas of the Americas, Russia, China, and India, many crops crucial to food supplies can no longer be grown. Harvests of wheat, corn, and soy beans are at record lows. Spring temperatures come earlier, disrupting planting seasons. Rainfall patterns have changed and much of what does grow is eaten by ravenous hordes of insects whose populations have been increased by the warmer temperatures. Even the famous Irish potato cannot be grown on the green island any more.*

GLOBAL POPULATION DECLINES: The number of people on Earth declined for the first time since the Black Death in 1400. Now however, there are almost 8 billion people, a number which is much more vulnerable due to overcrowding and diminished natural resources.

EARTH WILL BE UNABLE TO SUPPORT CIVILIZA-TION: Recent computer models predict that except in narrow zones near the poles and in high mountain areas, life for most people will become untenable within the next fifty years. Past predictions have been accurate but assumed a much longer time frame. Leading scientists agreed that when carbon levels reach 600 and average global temperatures are in the upper 60s, agriculture will be decimated, disease will be rampant and war will prevail.

Justification for doomsday predictions were bolstered from reports that thousands of plant and animal species that have existed for eons could no longer be found on Earth. Most of those listed were unknown to Stormy, but several stood out as EXTINCT. "Oh no, polar bears are gone," she suddenly realized. Those that remained on land after Arctic ice melted have all interbred with grizzly bears moving up from the south. As a species, they are gone. Orangutans, narwhals, and tigers have also disappeared.

"The canaries in the coal mine are telling us that our time is next," she told George.

"Polar bears may be the poster animals, but it is the tiny ones that make up the base of the food chains that are more important," he replied. "Things like ocean plankton that may not survive in a more acid ocean, or soil fungi that keep farmland fertile are essential to life on this planet. If we lose them, look out."

Confirmation of world events came from the United Nations survey of "Life on Earth" which concluded that thousands of plants and animals had not been able to survive climate change that now gripped the planet Earth. "Life as we know it is disappearing," she wrote.

Pierre Fights Back

Although he had not been seen in years, Uncle Pierre was apparently still active.

> *CLIMATE POSSE DESTROYS COAL PLANT: Reports from Wyoming indicate that the bombs thrown onto train cars full of coal were meant to damage electric power plants and thereby reduce carbon emissions. A shadowy group called The Climate Posse has taken credit for the devices, designed to look like lumps of coal but which are really explosives. Three coal generating stations have been destroyed and several more damaged. Local politicians have vowed revenge while green bloggers are cheering on the culprits.*

Lara - Born onto a different planet

It was 2058, three years after Jack's death that Lara came into the world. Stormy's granddaughter was beautiful, and for a while she erased the horrible memories of how TERRA had claimed Jack. Her daughter, Susana was an exceptional mother, just as Stormy had been, and the little girl was growing up strong and independent. Although she was heir to an Earth that was no longer benevolent to humanity, her birth marked a turning point.

"I feel like Lara will be part of a new era," Stormy told her daughter.

Sitting at the kitchen table, Susana looked puzzled. "Lara? A new era? What do you mean?"

"A period of great change, like we have every thirty or forty years when technology, social norms, and behaviors all change rapidly."

"Every thirty or forty years? Give me an example."

Stormy smiled and said, "if you remember your history lessons, the years around 1890 to 1900 were called the Gay Nineties. People started playing ragtime, railroads opened up the American west and women were starting to smoke and vote!"

"I remember hearing about those times," said Susana. "That was when electricity started to become available: Boy did that ever make a difference."

"Then came the "Roaring Twenties" when the stock market was making everyone rich, people were doing the Charleston dance and automobiles were popping up everywhere. Jazz came on the scene and art deco was popular. Telephones were invented too.

"Oh yeah, that's when horses and buggies went bye bye. What were the next eras?"

"The Sixties were next of course, and they were wild. Women got liberated, birth control became easy, and people were experimenting with new drugs. For most people it was a time when they could afford to buy a house and send their kids to college. You could see sex in the movies for the first time and the Civil Rights made a lot of progress. There was the anti-war movement and the start of environmentalism."

"Wow, that really was a period of change."

"There was kind of a lag then, but around 1990 – 2010, computers and the internet turned the world upside down. Information was rampant, people could carry a computer - phone everywhere and a lot of old industries like newspapers and buying music on tapes or CDs really suffered. Lots of new businesses started up though, and a lot of rich people got REALLY rich. Unfortunately, global warming began to really take off then too."

"So it's time for a new era," said Susana, but what will it be?"

"It's hard to recognize it while it's happening, but I think the average person (at least those who are not homeless, starving or fighting a war) is finally ready to make a sacrifice to save the planet. The technology has been here for a while and the old ways of making and using energy are dying." People realize that climate change can disrupt their food supply and the politicians are starting to follow their lead. Lara could be part of a new era which can lead to a more sustainable planet."

"Mom, you're usually right. And I hope you are this time too. Lara, you've got your work cut out for you."

<u>Civilization Fights Back</u>
After decades of indifference, Stormy's writings found an appreciative audience. With George's help, Congress approved massive investments in alternative energy. It had taken decades of debate, but the barriers to a long distance direct current and a more efficient electrical grid finally fell. New power lines to connect wind and solar farms began appearing on the landscape, bringing carbon free energy to masses of people in Chicago, Philadelphia, and Atlanta.

Also approved were strong regulations on the burning of fossil fuels. The use of coal, natural gas, oil, and even wood to produce electricity started to be phased out, year by year, until such uses were to be outlawed by 2075. Ration cards for such fuels were issued, just as they had been in World War II when supplies of rubber, sugar, copper and other essentials were in short supply. Riots ensued, but, as in the civil rights movement, the National Guard was called out to enforce the new laws.

As the production of fossil fuel fired electrical energy fell in America, Europe and then in China, greenhouse gases levels in the atmosphere began to level off for the first time in recorded human history.

Stormy celebrated with a cake, emblazoned with "Carbon Emissions Drop" written in chocolate syrup across the top. Her daughter, Susana, heartily approved, eating three pieces in one night.

TERRA - Unrelenting

TERRA, the Earth's climate, felt the downward pull as carbon emissions declined, easing the pressure to fill air with water and heat. But even as emissions slowly dropped while India, China and the United States built more solar panels and wind generators, methane levels rose dramatically.

Millions of acres of Arctic permafrost (including almost all of Alaska, more than half of Canada, and most of eastern Russia) were melting and thousands of acres of Indonesian peat bogs were being drained. All were giving up their stored concentrations of methane, the release of which could not be stopped. By 2060 methane levels were off the charts, more than making up for the drop in carbon. Many times more powerful in trapping the sun's radiation, the gas spurred TERRA on to even more deadly climate change.

As her average temperature increased from 61 to 62 degrees, she pushed sea level rise higher and higher, causing salt water to seep under dikes and contaminating coastal water supplies. In many coastal areas, millions were forced to buy fresh water to drink, but supplies were often not adequate for sanitation or industry.

She flooded coastal highways and cities, causing bridges and low areas to be abandoned and swelling the ranks of refugee populations. London's population declined forty percent.

Decimated by famine, drought, war, and economic death spirals that spun like massive hurricanes, her human population declined again, dropping to 7.4 billion. Not since the dark ages had so many lives been eliminated.

Stormy Dreams – Resolve

Restless in her bed, Stormy's dreams were invaded by a powerful presence. "I will destroy you," she heard as methane continued to bubble up from the Earth's vast tundra.

"You cannot stop me," continued the voice from the sky. "Storms, fire, floods, and drought are under my control."

Visions of massive clouds, surging waters, and unrelenting heat filled her sleeping mind while people scattered in fear before the raging climate.

As TERRA increased her grip on the earth, Stormy's unconscious began to fight back. "This will not stand," she moaned in her sleep. "It cannot be. I will stop it." Rising up above the Earth into the atmosphere, she stretched out her arms to stop the all-pervasive TERRA. Flashes of lightning flickered on her dark, bedroom walls as she shouted to a sea of humanity, imploring them to reduce the dangerous gases that were feeding an angry climate.

CHAPTER EIGHT – A CORRUPT ROBIN HOOD
2060 – 2070

Atmospheric Carbon Level - 510-535 ppm

Average Earth Temperature -62°
Sea Level Rise - + 6 feet
Earth's Population - 7.4 billion people

<u>Refugees</u>
Oooooo whaaaa hooooo. Oooooo whaaaaa hoooo. Wu hoo hoo hoo. The call of a loon ring tone erupted from her cell phone as a sleepy Stormy reached for it. It was Boothby.

"I need you in Arkansas," he said. "Cities down there are blockading the roads, trying to keep climate refugees out. There's been some violence, but most of the migrants are too sick to put up much of a fight and they just try to keep going. A lot of them have cholera, some have dengue fever, and of course they don't have health care and are homeless. Hundreds of them are camped out in the fields."

"I'm on it. I'll catch the local airline and report back to you tonight."

Stormy clutched her backpack as the small plane bounced in the wicked downdrafts of yet another big storm. Finally it began its approach to the Little Rock airport, made a jagged landing, and taxied to the terminal. Inside the concourse however, a voice on the loudspeaker told passengers to get back on board the plane.

"Attention. Attention: A quarantine is in effect. Under order of the governor, the airport is closed to all arriving passengers. You must get back on board the plane."

Complaining vigorously, Stormy stomped over to a very young man in a military uniform holding an outsized rifle. My God, he looks like an ROTC student she thought to herself.

"Press-WGMedia," she exclaimed.

"Sorry ma'am. I can't let you by: Captain's orders."

"You can't just close an airport," she insisted.

"You there; back to the plane," said the captain. "Private, escort this woman out of here."

"What is this? Military rule?"

"You're damn right. Now get moving or we'll move you. And put that camera away."

"Why are you doing this?" she demanded.

"Lady, the hospitals are overflowing, dead people are lying in the roads, nobody wants to touch them because they don't want to get sick, and farmers are shooting them in the fields. There are too many of them to feed and there is no vaccine for dengue fever. Most of the police force is sick, the mayor is dead, the National Guard is in the Arctic, and the whole city is under quarantine. Disease is following them wherever they go. It's the new black death."

Shaken, Stormy stepped back, thought a minute, and then headed toward the Ladies Room.

"The bathroom on the plane is out of order, so I'm just going to use this one," she told the young soldier. Ducking into a stall, she quickly hid her backpack behind the commode, thoroughly mussed up her hair and put on makeup, lots of makeup. Yuk, she thought to herself, looking in the mirror.

After reversing her jacket inside out and letting her blouse dangle below her belt, she stepped out of the room.

"You there, stop," shouted the guard.

"What? My husband's plane was turned around and now I have to go home alone," she replied. "You can't stop me from doing that, can you?"

"OK; back to the lobby," he replied.

Leaving the airport, Stormy caught a taxi and told the driver to head for the refugees.

"Lady, are you crazy? It's not safe out there."

"I know, and the rest of the country needs to know too. I'm a reporter," she said, handing him a $100 bill.

Ten miles from town they saw the first refugee camp. A white flag fluttered above one of the tents, haphazardly pitched in a barren field. To reduce the chance of catching one of the diseases they had, Stormy chose a lone individual and approached her.

"Hi, I'm a reporter for WGMedia. Can I talk to you?"

"Sure, don't worry, I'm not sick. At least I don't think so."

"Where are you from and why are you running?" she asked as they sat down in the shade of a large, old tree.

"I'm broke, I have no job, and the landlord kicked me out. The line at the food shelter is four blocks long and the Selma police told us to get out of town. I guess he thought I had the sickness or something," she said. "Where am I supposed to go?"

Stormy had no answer.

"What about the others, in the tents."

"They're too sick to walk any more. You don't want to go near them."

"I know. My parents died of Dengue fever."

"The cholera is worse. And most of them have not eaten in days. Do you have anything to eat?"

"Here, keep this water bottle," said Stormy as she handed her the little container. "I'll make sure you get some food. But aren't there local shelters or relief agencies in Alabama or Arkansas?"

"I don't know where you're from, but when most of the farms collapsed down here, everything went to hell. Our economy was based on agriculture and forestry, but now it's too hot for the soybeans and most of the trees burned up in fires. You can get food, but the prices are sky high and jobs? forget it. Half the population is looking for work. I tried going to the coast but the hurricane damage down there wiped them out too."

Seeing a police cruiser heading their way, Stormy quickly got into the waiting cab with her friend and they drove to the nearest restaurant. There, after a dinner of pasta and salad, Stormy offered her a computer phone.

"I've seen those, but I don't know how to use one. My family has been dirt poor for years. You know what I mean?"

A tear dripped down Stormy's eye as she realized she couldn't solve the economic crisis, the famine, the refugee crisis, or the climate change that spawned them. "OK, but take this money," she said, and good luck. I'll see what I can do in Washington."

Back at the airport, Stormy realized that Houston, Memphis, and most of the cities in Mississippi were also establishing quarantines and turning away planes. Apparently there were other bands of people displaced by the famine, forest fires, and six hurricanes that have slammed the

Gulf Coast that summer. Reluctantly, she rented a car and drove back to Washington.

Back in the office, Stormy began to piece together a report on the quarantines that were sweeping the south. At first she began to write about how local bullies were usurping people's rights, but as she heard from more and more cities and towns, she realized that officials had no choice. Weakened by an unemployment level of 25-50 percent, sky-high prices for gasoline, and massive numbers of people with asthma, malnourishment, and contagious diseases, they had gone into a survival mode.

Worldwide, the situation was worse, much worse. Although numbers were uncertain, estimates by military agencies indicated that more than fifteen percent of Earth's people were now refugees, displaced from their homes without food, water, shelter, or hope.

And Now the News: Dateline 2064

SEA LEVEL RISE REACHES 6 FEET: Exceeding previous estimates by several times, global sea level rise reached six feet this year. Scientists blamed the increase on faster than expected warming of the ocean and the continuing collapse of ice fields in Greenland and Antarctica.

MASSIVE FLOODS CAUSE DAM FAILURES IN A DOZEN STATES: Old dams failed on the Shenandoah, Chattahoochee, Hudson, and dozens of other rivers last year as floods expected once in 200 years breached spillways and washed out foundations. Downstream losses were estimated in the billions of dollars and hundreds lost their lives. Along the Missouri and Mississippi rivers, the annual flood intensified, washing millions of corn stalks down into the Gulf of Mexico, feeding the growing dead zone.

MASSIVE FLOODING IN CALIFORNIA: Politicians are fleeing the state capital as water pours in through levees damaged by the recent earthquake. Sacramento became increasingly vulnerable as sea levels rose, backing water up into the delta on which the old city had been built. In the face of continuing ocean rise, cost estimates to repair earthquake damage while doubling the size of the levees and rebuilding the city were prohibitive according to the Army Corps of Engineers. State officials agreed to move the capital to higher ground and convert thousands of acres of the former city into an aquatic park. Efforts to block the tide at the Golden Gate Bridge with a massive dam were rejected due to concerns over vulnerability to terrorist attacks. "If we built it and they bombed it, the flooding would make Hurricane Katrina damage look like child's play," said Boothby.

SIBERIA, CANADA ESTABLISH NEW "CLIMATE OASIS" CITIES: Driven by deteriorating conditions to the south, northern nations have created new cities for the wealthy in areas beyond the reach of global warming. Demand is outstripping supply as people abandon coastal areas and expanding deserts.

FIRST CATEGORY 7 HURRICANE CRUSHES GULF COAST: Striking New Orleans again as she did in 2005, a hurricane named Katrina II flooded much of the old city, breaching sections of dikes not previously upgraded. Hundreds of the 7,000 offshore oil platforms have been destroyed and massive oil slicks have coalesced into one enormous mass, covering one tenth of the Gulf of Mexico.

AMAZON RAIN FOREST BURNS: Satellite images revealed today that enormous fires have engulfed much of the Amazon rain forest. Only a few years of drought are required to create tinder dry conditions reported local researchers. Flames spread as thousands of square miles continued to burn.

AMERICAN, SOVIET, AND CHINESE GRAIN BELTS WILT: With ground water levels too low to economically pump, wheat and corn production has declined 80 percent. Estimates are that half of the world's population now goes to bed hungry every night. By 2100, average summer temperatures in the American mid-west are predicted to be hotter than the hottest summer days a century ago.

ITERNATIONAL AID HALTED: The United States and other countries have stopped sending food and aid to impoverished nations reported the UN today. "Local conditions demand that we help our people first," explained the Secretary of State.

CLIMATE POSSE STRIKES AGAIN: Emboldened by their success in crippling coal generating plants, angry activists have expanded their targets to include cement plants, oil refineries, and coal mining operations. Dozens of such facilities have been bombed by what appear to be numerous and growing cells of climate anarchists. In a particularly brazen attack, they devastated the Keystone pipeline, bombing several pumping stations along the route in Nebraska and Texas. "Kill a coal plant: Save the Earth" was emblazoned across their posters.

Lara Dreams

"Mommy, help!" cried Lara.

"I'm here. What's wrong?"

"There was a big white flash and then everything was gone," said the tiny girl. "Then I woke up."

"Go back to sleep darling. Try to dream about nice things."

Stormy Reports from Africa

A rocket-propelled grenade was pointed at the windshield. Two men in ragged clothing on each side of the bus trained assault rifles on the frightened passengers.

"Get down. Everyone off the bus," shouted the terrorist.

On a world tour of climate change, Stormy was riding with Sahela, her new friend to report on the desperation of a people devastated by the convergence of poverty, political corruption, depletion of natural resources, too many people, and now, rainfall patterns that no longer supported farming. Their driver had failed to pay for an armed guard, who had called ahead to notify his companions of the oncoming bus.

Sahela was born in Africa into a life very different from Stormy. Decades ago, her people had survived by farming while their neighbors herded cattle to areas where seasonal rains fell. While Stormy had ridden in comfortable air-conditioned cars, Sahela had walked dusty roads, carrying water. While Stormy had played computer games and attended a university, Sahela had hoarded grain and read scarce books by firelight. While Stormy's culture emitted the greenhouse gases that were changing climate, Sahela struggled to cope with rains that fell at the wrong time, or all at once, or not at all.

"Give them what they want," said Sahela, "and don't make eye contact."

But Stormy did look at them, noting their sunken faces, ragged clothing, and gaunt bodies. These people were serious, and her death would mean little to them.

Their leader, a young man of about seventeen, quickly collected food, water, and anything of wealth from everyone. But it was not enough. Grabbing Sahela's daughter, they quickly vanished in the dry scrub land.

Screams, panic, anger, and then despair gripped Sahela as she struggled to comprehend the situation.

"They will want a ransom," she declared. "We can find out how much from the local police because they will get a share of it."

Suddenly, Stormy realized she was the only wealthy person within miles. Surrounding her were members of an ancient African people living in poverty, despair, and increasing hopelessness as shifting climate zones continued to intensify the now decades-long drought. These people are hungry, she realized, on the verge of starvation. Society has dissolved into a morass that pits neighbor against neighbor, friend against friend, and haves against have nots. But I am surrounded by good people, she thought. Good people whose lives have been determined by changes beyond their control. Born onto a planet whose atmosphere is increasingly laden with gases that trap heat, they are victims of lifestyles in faraway places. Places like New York, Peking, and Calcutta where carbon emissions carry shoppers to stores, cool houses, and build ever larger edifices to wealth."

"I can pay the ransom," said Stormy. "We've got to get your daughter back."

As night fell they were led, blindfolded, to a primitive hut somewhere beyond the lights of town. There they confronted their captors and negotiations began.

"We want one million American dollars." said an elderly man, firelight flickering from the barrel of the old soviet rifle he was holding.

"I cannot pay that, but..."

Stormy's words were interrupted as an angry fist fell upon her cheek, leaving a cut and later a scar that would hurt whenever TERRA's winds carried a low-pressure system.

"You Americans: You're all alike. Filthy rich but you don't care about my starving children!" shouted her attacker, a young man dressed in a ragged Coca Cola T-shirt.

"Leave us," said the old man, obviously their leader, as he pushed the youth away.

Regaining her composure, Stormy continued, "I will give you one thousand dollars, and I will make certain that a new water well is drilled in your village and provide you with medicine and food," replied Stormy.

Her response froze the old man for an instant. Their eyes locked.

"You can do this?" he exclaimed.

"Yes, I personally know people in the local relief agency, and if you give up your guns and stop the raids, we can provide tools to plant trees."

"We will never give up our guns, but the well... we need the water. Bring the girl out."

Tensions cooled as Sahela was reunited with her daughter and Stormy offered the sack of food she had brought along to share. They were astonished at this brash white woman who stood up to their guns, yet wanted to be their friend. Stormy broke the bread and gave it to Moisha, a thin tribesman who put down his rifle and quickly devoured the scant meal, his first of that day.

"Tell me why you steal," said Stormy.

"We have nothing. Our cattle and goats are dead; there is no grass for them to eat. Other tribes raid us and locusts eat what we try to grow. They cut my wife's throat and took my children while I tried to defend our village. The land has become a desert, there is no government, and we have nowhere to go."

"I understand," said Stormy, as she held out her hand. "Tomorrow I will return to Nairobi and arrange for a new well to be drilled here, and a wind generator to power the pump. Then you must plant trees: Trees to shade the soil and help retain moisture. A local food truck should be here next week. And Moisha, maybe you should not have any more babies until there is enough food to feed them. Learn how to do this and your people will be better."

Under a spinning fan in her shabby hotel room, Stormy transmitted her report to WGMedia.

"It is a world of utter despair, dominated by pestilence, war, famine, and death. To these have been added the fifth horseman of the apocalypse, climate change."

"I cried when I saw the starving children," her article continued, "their distended bellies and fearful faces crying in the dust. I gave them all the money I had, but later realized that it would only alleviate their suffering for a few days. Indeed, money cannot change the shifting winds that now carry rainfall out to the oceans where decades ago it fell on fertile fields.

As famine, violence and war stalked Africa, Stormy returned to her desk in America, more determined than ever to do all she could to stop the emissions that were feeding the convergence of poverty, resource degradation, and climate change.

Exego's Plan

"We can't print this; they would deny it," said Stormy as she walked into Boothby's office, turned on her cell phone, and placed it on his desk. She had that mischievous look on her face.

"Why can't we print it? This is a free country."

"They didn't know my phone was turned on and recording," she answered, obviously proud of herself.

The interview had been with executives of Exego, now the dominant oil company in the United States. Gas prices had leaped again and she wanted to know what they would do now that Middle East oil was being sent to China, India, and Japan. The meeting had been drab; the usual self- serving answers, justification for more record profits and how they were going to finally get the oil under the Arctic.

Ending the bland conversation, Stormy *accidentally* left her cell phone on the chair, thanked the suits, and walked out. But the little machine, returned intact to her later, yielded insights to the power structure of the world. Playing it back, Boothby heard:

"The bitch is gone."

"I don't trust her."

"Yeah but she can't stop us. Let's get going on this Arctic drilling. I want that rig up next month."

"It will cost us, 5 billion at least."

"So, what?"

"What about the other countries? They claim that territory."

"Fuck 'em. If we put in a rig, all they can do is complain to the Arctic Council and we control the votes there. Old Lars, the Norwegian representative, is in love, or should I say lust, with Janah and she'll, how shall I say it, *convince him* to vote our way."

Janah was an asset, and on the payroll of Exego. Young, extremely attractive, and a graduate of both modeling and acting academies, she could persuade any man to do almost anything. Over the years she had delivered millions of dollars in oil contracts.

"Janah. Oh yeah. When is she coming by here?"

"She's out of your league. I had her once; thought I was going to have a heart attack."

Stormy's illicit recording continued.

"There's no worry at the Supreme Court. Cedric can swing the votes there."

"Go ahead and build it. There are billions and billions of dollars down there. The Russians are too chicken to attack, but if we don't get there first, they might start drilling first. Gentlemen, this is our chance to become millionaires, multi-millionaires."

"Okay, it's settled. Call in the usual contractors and get her built so I can have gas for my SUV."

"Come on Frank, you know it's free for us guys. We don't pay for our own product."

"But what about the environmental regulations for the drilling station?"

"What about 'em?"

At that point the recording went dead, ended as unknowing participants discovered the phone and turned it off.

"Stormy, you are unbelievable," said Boothby. "Nice work."

"Well, at least we know what's coming," she replied.

Hurricane Walter Hits Washington

It was 2066, and a category 5 hurricane had never hit Washington D.C., but in September, it did. Roaring up the Chesapeake Bay, the monstrous storm gained strength from the hot, shallow waters, slammed into the nation's capital, and then moved on into Pennsylvania.

Shortly afterwards, climate refugees massed at the Capital mall, demanding food, shelter, and clean water. George wanted to talk to them; to tell them of the bill he had introduced on their behalf.

"I've got to get down there," he said.

Stormy begged him not to go. "That ill-intentioned wind just might know your name," she said. "The news just reported that the Washington

Monument has blown over; it's laying on its side in pieces. There are tornadoes in the area.

"Oh you've been listening to Tom Russell again, haven't you?" came the reply. "I have to go. People need to be told that this kind of a storm is our future unless we get control of green-house gases," he said, as the screen door slammed in the wind.

Strong winds pushed his little car around as he sped down the beltway, but the road ahead was blocked by water several feet deep. It must be a plugged storm drain, he thought, as he turned off onto side streets. Slowing to avoid broken branches in the street, he stopped too long. The tornado was approaching, and the 175-mile-per-hour winds were too much for the big maple tree. Down it came, crushing the car and trapping George within it. George gasped as the tree limb, more than a foot in diameter, fell across his chest, pinning him to the collapsed seat. Grabbing his cell phone, he quickly called Stormy for help.

But the only sound he heard was TERRA, shrieking static into the speaker.

His death was violent at first, then quiet as internal bleeding drained the life away from his body. He sat alone in the car while more and more of his blood flowed through a broken artery into his stomach. A warm feeling of delirium enveloped him and as he slipped into a stupor, mumbling about his love for Stormy. Climate had claimed another victim.

At his funeral, a tearful Stormy remembered the words of her friend.

> *"The world's not round now, it's not even flat;*
> *It's crooked and craggy like Mount Ararat;*
> *Been squeezed out of shape in the madman's grip,*
> *So no one can make it to the mythical ship.*
>
> *So I'm going down under, to the bottom of the sea*
> *To float with the fishes, they don't bother me.*
> *Rather drift with the currents, make food for the sharks,*
> *Than to live with these humans,*
> *That are breaking my heart.*
>
> *Midnight and downers, hot milk you can keep,*

If I could have someone, to love me to sleep."
Patricia M. Long/aka Patricia Hardin.

Devastated, Stormy sought refuge in Canada with Susana and Lara, her granddaughter.

"It's as if all the life has drained out of me," Stormy said. "Maybe I'm like this old planet. The structure is there; the cities, buildings, people and my body, but it's all changed. The sun, earth, water, and sky are still here, but life itself is different. The seasons, life zones, weather patterns; they're all changed. And like death, we can't do a damn thing about it."

But life for Stormy did continue. After three months of contemplation, she returned to Washington and resumed her work.

Conversations in a pub

Candles lit the bar as Stormy and Boothby ducked in from the howling wind.

"What happened to the power?" Stormy asked.

"Lightning knocked out the power: I've never seen so many flashes," said the bartender, "dozens every minute." The whole substation blew up last week and they can't get parts because the same thing is happening everywhere else. What's going on?"

"It's simple physics," she replied. "When the air gets warmer, it holds more water. And since heat rises, it makes thunderstorms. The hotter the air, the more the rain, and what goes up, must come down, along with the lightning."

"Those damned laws of physics, why doesn't somebody change them?" he asked.

Stupor was a new bartender. He was young and, like many others in his generation, he had skipped school to focus on paying the rent. Knowing that it would be useless to continue the conversation, Stormy ordered a drink.

"That new wine from England is great. Can I have a glass?"

"They don't grow wine in England," said Boothby.

"They do now. Too bad about Spain though; it's too hot for the grapes there now."

"And here's to those poor farmers in California," said Boothby as he raised his glass. "Thousands of them left the valley when it dried up. I guess thirty years of drought was too much."

"I don't believe that rubbish," said Stupor.

"Well you better," came a voice from the bar. "I used to live in Fresno, until unemployment hit 50 percent. You can't grow crops without water."

The stranger startled them.

"Come on over," said Boothby. "I'll buy you a drink."

"Boothby!" hissed Stormy. "We don't know this guy."

"There's a story here," he replied.

Juan had lived in California for forty years. The son of an immigrant, he had worked his way up from picking cotton to growing oranges and almonds. But as snow packs that had supplied irrigation water diminished year after year, reservoirs became inadequate to hold enough water in the summer for the massive farms. The final blow came when levees succumbed to sea-level rise and salt water from the ocean flooded the Delta and then the aqueducts. Year by year the farming economy crumbled, and with it the lives of people who had depended on it. Migration from the San Joaquin valley was in full swing and families from California were showing up looking for work in every state east of the Mississippi river. Jobs were not available in the southwest, which was becoming a dust bowl as rain patterns shifted northward. Nor were they available in the Pacific Northwest where bark beetles had killed millions of acres of trees. They were not welcome in the plains states either, where immigrants from Mexico were moving in. Some had found work fighting the perennial wildfires, but most, like Juan, had just kept moving.

"I held on as long as I could," said the grizzled farmer. The lines on his face seemed to grow deeper as he remembered his former life. "I had a family. We didn't have much, but there was food on the table, a small house and love, lots of love. Oh God my wife was beautiful. Juanita, my daughter, was going to start her college studies. We had a kitten. Then, when the big rains came, the levees broke and the flood took everything away. Our house was ruined. There was no land to work and the boss fired everybody. My wife got the dengue fever from all the mosquitos and died. We had no money so the doctors wouldn't help us. It was horrible. It was too much for Juanita and she went to a commune in the hills. I don't know where she is now."

The old man had tears in his eyes.

"It wasn't your fault," said Stormy, reaching out to hold his hand. No one can stop these climate changes. It wasn't your fault. That's an amazing story. I'd like to add this to my blog if you don't mind."

"Oh sure: I've read a lot of your work," he replied, brightening up a bit.

" In fact, many of us climate refugees want you to run for president. If George could do it, you can!"

The idea stunned Stormy, but remained with her and memories of Goethe's poem filled her dreams.

*"Whatever you will, or dream you can, begin it.
Boldness has genius, power, and magic in it."*

Bolting awake, she wondered: *Who is bolder, TERRA or civilization? Who is more powerful, the forces of climate, or people?* It was a question she pondered as she considered running for office.

A Run for the Presidency

Glittering lights filled the ballroom. People were dressed to the nines and tuxedo-clad waiters kept asking Stormy if she wanted more hors d'oeuvres. Although George was gone she stayed in contact with many of his colleagues, so when the opportunity arose at a cocktail party, she approached Senator Porter who was retiring from his seat in Illinois.

"Senator, what do we need to do to solve this climate crisis?"

"You should run for my job," he said. "You still have a residence there and people know you. Not only that, you're very popular. Everyone has heard of Stormy the climate crusader. Having no ties to an established party and memories of how hard George worked will get you a lot of votes, too."

Stormy thought long and hard about it, and when Boothby began encouraging her, she agreed. She would run for the Senate, then for the presidency. There had not been a president from Illinois since Obama, whose efforts at slowing climate change had been weak and were blocked by radical conservatives and over the top enviros. But first, she would start a new political party.

And so was born *The American Party*, a coalition of everyday people committed to reversing climate change, the re- ascendance of science, and the restoration of common values. For her campaign staff she recruited a baseball hero, a musician from the pop charts, a movie star, and a late night talk show host; folks that could relate to everyone, not just to money.

As she contemplated her actions, Stormy remembered the words of Abraham Lincoln when he confronted his great challenge. Speaking into the computer, she began to recite.

"Two hundred years ago our civilization embarked on a new way of life, dedicated to the proposition that we could burn as much fossil fuel as we wanted. Now we are engaged in a great planetary crisis, testing whether our people, or any people so dedicated and so selfish, can long endure.

"We are met on many battlefields. We are assaulted by famine, by drought and by our own people as we struggle to find common cause with the earth. Our enemies are greenhouse gases released from melting permafrost and destruction of forests, the hurricanes that assault our coastlines, the raging floods, region wide storms, and indeed, our very selves.

"Our jet streams have shifted, our sea levels are rising, and our deserts are spreading. As they do, the wealthy are fleeing the ravages of collapsed economies, of disease, and of our dislocated citizens: Fleeing to protected areas, secure in the profits of their despoliation.

"But in a larger sense, can we control not only our self destiny, but also the Earth? Can we reverse the centuries of technology gone astray? We can. We must; for our very lives depend upon it. We do this not only for ourselves, but also for our children, their children, and their children's children; the people and the life of this planet.

"On this day in 2068 we resolve that the policies of the past, the nationalism and discounting of environmental costs which have brought our civilization to the brink, shall be changed: That this world shall henceforth be dedicated to controlling

climate destruction; and that the people, and all other forms of life, shall not perish from the earth."

"Vote for Stormy and The American Party!" yelled Boothby, Susana, and Lara at rallies on Chicago's State Street, in Springfield and even at small towns like Shabbona, Illinois.

"Vote for grandma," echoed little Lara.

"Vote for mom. Climate stabilization will mean better apple pies," wrote Susana in her new job as a reporter.

"Vote for me," appealed Stormy, "and I'll put a solar panel on every roof. With your help, we can beat this global warming monster."

Shaking hands with supporters at rallies, Stormy listened to their comments.

"I'm just trying to survive this heat wave," said a young woman. "Do you know where I can buy an air conditioner?"

"I'm sick of these power black outs," said the lady in red. "Even food in the stores is spoiling."

The message resounded, found fertile political soil, and grew. In twelve short months, Stormy sat in what had been George's chair in the United States Senate. Efforts by the old climate denier coalition to stop her could go nowhere in the face of near continuous storms, unpredictable weather patterns, and a sputtering agricultural economy. Six years later The American Party swept her into the office of the presidency.

The new political party's efforts bore fruit when, in an act unbelievable just 50 years ago, the use of coal became illegal in the United States and elsewhere, including China. Carbon police were created to enforce provisions of the "World Security Act" as written by the United Nations.

In America, Europe and China, special branches of the military were created to insure compliance. Punishments for gross climate pollution became severe.

"As of this day, fossil fuel use for generating energy is hereby nationalized," proclaimed Stormy in response to a referendum of the people. Riots ensued as coal and gas fired power plants were forced to close.

In a final blow to Cedric and the climate deniers, the law (which passed with 75 percent of the votes in Congress) penalized lies, deceit, and corruption regarding climate change in publicly broadcast advertisements and similar media. Anyone could still say anything they wanted to (and some still blogged about how current disasters were part of natural cycles) but they would now have to answer in court to prove their allegations. If they could not convince a judge or panel of certified scientists, jail time was meted out, along with stiff monetary fines. Public opinion polls reported that a vast majority of the people supported the new law.

"You can't legally yell *fire* in a crowded theater," said a man on the street, "and you shouldn't be allowed to put people in jeopardy by misleading them about climate change either."

"In America, we don't let liars, lobbyists, and big corporations determine public policy with unproven allegations," said Stormy as she signed the law. "People are sick of being deliberately misled by those that seek to profit by using the air as a sewer for their greenhouse gases. Years ago we exposed their tobacco fraud, and now we are exposing the clean coal fraud. If you say it, prove it: It's as simple as that."

Lara was at the ceremony and collected three of the pens used by her grandmother to add her signature to those of other world leaders. Reporters could not help noting that the brash little girl bore a striking resemblance to Stormy.

After passage of the "Climate Emergency Act" factories for wind generators and solar panels sprouted in formerly abandoned industrial districts. Environmental concerns over wind generator bird strikes and power line invasion of wildlife habitat vanished as even the most ardent of avian activists realized that climate change was adding another age of extinction: The Anthropocene.

After years of delay, China, the United States, and the European Union finally began building fourth generation nuclear plants to provide power formerly produced from coal, oil, and natural gas. Using radioactive waste from earlier nuclear plants, these new plants would eventually eliminate the terrorist threat of converting such waste to bombs. Built along major rivers in areas far from flood plains and sea level rise, they slowly began to replace the electrical power lost when fossil fuel plants were banned.

Across the nations of the world, thousands of bio-char factories were established to partially burn organic matter, turning carbon into a charcoal like substance, impounding what now was becoming known as the villainous gas. The end product was char, which could be buried in soils to increase their fertility and help feed a hungry world. It was a win-win, low-tech answer to climate change that could be easily implemented. "We should have done this on a massive scale decades ago," said Stormy in her press release.

For a brief moment in time, and for the first time in human history, the world was united as Stormy rallied other nations to join in the fight against climate change. It was a fight to reclaim their planet, and their destiny. "Today we stand together; all races, all ages, all people, to work for a common goal," proclaimed Stormy. "Carbon emissions must fall, and they will; they will."

But as methane levels increased, TERRA continued her rage.

Senator Inhuff

Reading an assignment on her tablet, Lara saw a photo of an old U.S. Senator who had called global warming a hoax. "Who was Senator Ina-huff?" she asked.

"He was a real jackass," replied Stormy. "Your grandfather once threw a cream pie in his face, and the whole United States Senate cheered."

"What's a jackass?" said Lara and she went to look it up on the web.

Definition: *Jackass: a man who is a stupid, incompetent fool.*

Reading about his career as a denier, she blurted out "that guy was a real BIG jackass. But why did grandpa throw a pie in his face?"

"Senator Inhuff said that God was in charge of the climate and that the Senate shouldn't interfere with whatever God does. So your grandfather told him that America was based on separation of church and state, which means that religious beliefs should not interfere with the laws that are passed. Then Senator Inhuff said that God might strike him down for saying that, so grandpa hit him in the face with the cream pie that I gave him for lunch."

"What a waste of your pie!" said Lara. "But why did some people believe the Ina-huff?"

"Well, I'm reading this book by a man named Curt Vonnegut who was writing about climate change, and he said:"

> *"We probably could have saved ourselves, but were too damned lazy to try very hard... and too damned cheap."*

"When I grow up, I'm going to be a senator, just like grandpa was," Lara said.

"I hope you do," said Stormy.

Lara dies

"Look, a butterfly!" said Lara.

Running across the Nebraskan prairie, little Lara was eleven years old and full of energy. Vacationing with Susana and grandma, she could not be restrained, so they let her play in the fields of grass and flowers. Overhead, the sky seemed to surround them on the treeless plain.

"Susana, your fried chicken is fantastic, but I can't eat any more," exclaimed Stormy.

"It's just as well," came the reply as her daughter noticed the rapidly darkening sky to the west. "We had better get back to the car."

A sudden wind blew across the picnic blanket and to the north, lightning flashed.

"Lara, come back. It's starting to rain."

Seemingly from nowhere the black clouds emerged, following them to the road.

"Where's Lara?" said Susana.

"Oh my God, she is still back there getting her teddy bear."

Without warning, without malice but with deadly accuracy, 100 million volts of lightning struck the little bear just as Lara reached for it. A thunderous roar and blinding light obliterated the scene, followed by an instant of silence that filled the air as their eyes and ears recovered from the shock.

"NO!" screamed Susana when she saw Lara motionless on the ground.

Instinctively, Stormy held Susana, shielding her eyes from the smoking teddy bear. Remembering that lightning victims can sometimes be revived, she began to pump her tiny chest and breathe into her mouth. But TERRA had been too strong for the little body and Lara was dead.

The storm moved quickly on, flashing its light to the south; leaving them alone on the prairie. On that day, tens of thousands of lightning strikes, often dozens per minute were recorded by the satellite overhead. Heat, moisture, storms, and lightning; they had all multiplied as the climate increased its grip on the planet.

In her fireside chat, Stormy addressed the nation about the climate disruption that has taken her parents, her son, her husband, and now her granddaughter:

"There is no enemy, because the atmosphere is everywhere. There is no target to attack. At this point, there is no way to win. It is not like an athletic competition, or even a war. It is only a slow, hundred-year deterioration, and we don't seem to be able to do anything about it. Instead, we keep blaming each other. And people are dying.

In the beginning, it would have only required a token sacrifice from each of us to prevent the massive buildup of greenhouse gases. But as the decades passed, greater changes were required, but not made. Now, the price has become more than most can bear, and it is being paid by those of us who have lost loved ones. Good bye, Lara. Stormy loved you."

TERRA – Earth Succumbs to the Sea

Driven to expand by warmer temperatures and melting ice, TERRA's oceans had reached 7 feet above what human civilization had ever experienced. Satellites whirling above her sent back images from around the planet: Miami, a city built on Swiss cheese like porous rock, had been abandoned. Everglades National Park was now a marine reserve. Shanghai and Mumbai were becoming islands. Venice, Amsterdam, and New Orleans were going underwater, as was the northern third of Jakarta. Vast areas around Hong Kong and Tokyo were flooded. Most of the Florida Keys and much of Cape Canaveral were about to vanish and if it were not for miles of protective dikes, salt water would be approaching the White House.

Flood control structures protected other areas, including ports, harbors, and naval bases around the world, but continual maintenance in the face of TERRA's rising seas had become more costly than many governments could afford. Quietly the levees waited for the next storm that would breach them and free the lapping waters to roam streets, fill houses, and drive people away.

Stormy dreams - Resentment

In her darkened bedroom, Stormy clutched a pillow, fighting the images that filled her sleeping mind. An ill wind blew aside the curtain and wrapped itself around her. "It should not be this way. I am not at fault and I have tried, over and over, to change it. I hate this climate and I hate the damn politicians who have ignored it to feather their own nest. They talk of evil taxes while seas rise and people die."

"TERRA is a bitch. No one can reason with her. Everything I have done has been in vain and I cannot go on, but I must. I will."

When she gradually awoke in her White House bedroom, she realized that the dreams, and the emotions she felt in it, were real. As the skies grew ever darker, so did her inner thoughts. How much worse can it get, she wondered?

CHAPTER NINE – ACCELERATION
2070 – 2080

Atmospheric Carbon Level – 535-560 ppm

Average Earth Temperature -63.5°
Sea Level Rise - + 8.5 feet
Earth's Population - 6.7 billion people

<u>Don't Wake Up</u>
She didn't want to wake up, but the president cannot oversleep. There were people to meet, bills to sign, press conferences to attend.

"Mrs. President: Do you support National Manufacturer's Day?.... Mrs. President?"

Stormy was day dreaming, repeating the nightmares of early morning.

"Manufacturers? Why yes; of course we need more: Next question."

But as the conference droned on, her attention kept wandering back to the Canadian Inuit who had implored her to save their people and a way of life being torn apart by melting ice, changing weather, and the devastation of their culture. Desperate, many were actually eating their sled dogs that were of no use when there was no snow.

"What about the proposed initiative for deleting the Jefferson nickel from the currency asked another reporter? The inflation rate has made it useless."

But Stormy was gone, her mind having drifted to the lost nations of flooded islands in the south Pacific.

"Mrs. President. Are you all right?"

Snapping to attention, she righted the ship with an agreement that the nickel must go, but then proceeded to draw a parallel to the loss of indigenous cultures.

"Thomas Jefferson died; and the nickel will be gone, having outlived its usefulness. But the people, people just like you and me, who have been forced from their ancestral homes in the arctic, from their lands on the low lying islands and from the slums of Bangladesh, did not have a choice. We; you, I, and all those who have been spewing billions of tons of carbon into the air for a century, have taken away their culture. Hundreds of thousands have been annihilated. Millions more have become climate refugees, searching for a place to live, a place to survive without hunger, war, or death. And the world has done little for these people. I suppose it was like that during World War II when the Nazi's exterminated the Jews. Like the holocaust, climate crimes of such immensity must never be forgotten."

Realizing that the room has gone silent, Stormy could only mumble the thought that had pervaded her thoughts for the last year. "There can be no freedom under an oppressive sky," she said. "We no longer have the freedom to act as if our collective behavior does not change the planet."

Silenced and embarrassed, reporters drifted out as the news conference ended.

Goodbye My Friend

New Year celebrations were few and far between as the seventh decade of the twenty-first century ended. But that tradition was not the only thing that was ending. As the New Year began, Boothby died. Recently retired, he had moved to the eastern shore of Maryland. There, in a cottage by the bay he was one of several hundred elderly souls that succumbed to the searing month- long heat wave. Electric power for air conditioning was very intermittent during such situations as power demand spiked beyond limits of the local, antiquated grid. It was not so much the month-long 110-degree days as it was the hot, 85-degree nights that sapped the strength of everyone.

His life had been one of endless searching for news, continual astonishment at what he found, and warm relationships with those he knew. Born into a world that would not recognize the danger of climate de-stabilization, he had watched TERRA's power grow, slowly at first, and then become unstoppable as humanity continued to feed the greenhouse gases that increased her strength.

The funeral was simple, as he wanted. But along with the ashes that Stormy retrieved from the crematorium was a note containing Boothby's last wish. "Sprinkle my bones between the cracks beneath the bar at the pub," he wrote. "Then I can continue to hear the debates, the joys, and sorrows that take place there. And maybe some Irish whiskey will drip down on me once in a while."

Stormy complied with his request, forcing back tears as she said goodbye to her mentor. TERRA had claimed her best friend and now only Susana was left. Together they faced a changed and changing world.

Desperation

Society's efforts and the Climate Emergency Act made progress as world carbon emissions declined, but feedback loops triggered by decades of indulgence continued. As the great warming increased, humanity became even more desperate.

"Mrs. President, I have terrible news."

Hans was her climate administrator, and he was not prone to such outbursts.

"Stop trembling and show me the photographs." said Stormy as she got up from the oval office desk. "What are these images?"

"They are photos of Venus, made with the new radar imaging satellite. It can peer through the dense atmosphere there and as you can see, there are ancient structures on the surface of the planet!"

"What are these?" asked Stormy. Her brow became furrowed as her eyes narrowed.

"Some kind of transportation network," he replied. "And look, those were buildings, lot of them."

"My God, there was a civilization there. That is truly astonishing. Has NASA confirmed this?"

"Yes, an entire planet that once supported an advanced civilization is now so hot that no known form of life can live there. Even lead melts on the surface of the dark side."

"Carbon levels on Venus are over 900,000," replied Stormy as she sank into her chair. "They must have had runaway climate change that burned all of their organic matter into the atmosphere. How high would carbon levels have to be on Earth to wipe us out?"

"We don't know, but if our fossil fuel deposits are burned up and ocean methane deposits are released, average global temperatures would be in the 70s, we would be looking at an eventual sea-level rise of over 200 feet and most of the people on Earth would be starving, disease ridden refugees. That would pretty well knock out civilization as we know it."

"If it happened on Venus, it could happen here," murmured Stormy.

"We are losing the battle," said Hans. "We've got to do some-thing different, and do it now, or it won't matter."

In what she described as a last gasp effort, Stormy turned to geo-engineering. They would block sunlight with sulfur spread high in the stratosphere. Jet fuel was soon laced with sea salt and sulfur to carry the reflective chemicals into the stratosphere.

Walking in a park by the mall an elderly woman heard her daughter say:

"Look mom, up there."

A yellow green color infected the sky, put there by hundreds of jet planes spraying sulfur ten miles above their head, a form of planetary chemotherapy designed to block sunlight, but which had other, poisonous affects. It was a desperate attempt by mankind to cool the runaway climate. In the end, hydrogen sulfide from the ill-fated attempt created critical holes in the ozone layer and skin cancer blossomed on human flesh below. The effort, only mildly successful, was discontinued and temperatures continued to rise.

As the health of human civilization declined, so did democracy. Three, four, and then five new political parties sprang up, but as each new politician failed at stopping climate change, fewer and fewer people voted. Political processes were not very good at taking money from the current generation to save the next one, and those that tried were voted out of office.

Discouraged, Stormy refused to run for re-election and retired.

Explosions in the Arctic
Ice kept forming on the spinning drill as workers struggled to remove it. Twenty eight Exego employees lived on the stadium size Arctic oil platform built in waters twenty miles offshore of the Northwest Passage. A dim light glowed from the galley in the pre-dawn stillness where men sipped coffee and dreamed of liquid gold. Five thousand feet below them lay a wealth of oil, waiting as it had done for millions of years since being created from carboniferous ferns and bodies of giant reptiles. Oil that would be used for gas tanks of SUV's and profits of silk-coated executives.

Or would it?

Five thousand feet above them the warhead began its descent. Heat from friction with the carbon-laden atmosphere caused its nosecone to glow red, illuminating the serial number USA 5302N. This particular missile had a long history. Built at a secret facility, someone had duplicated the serial

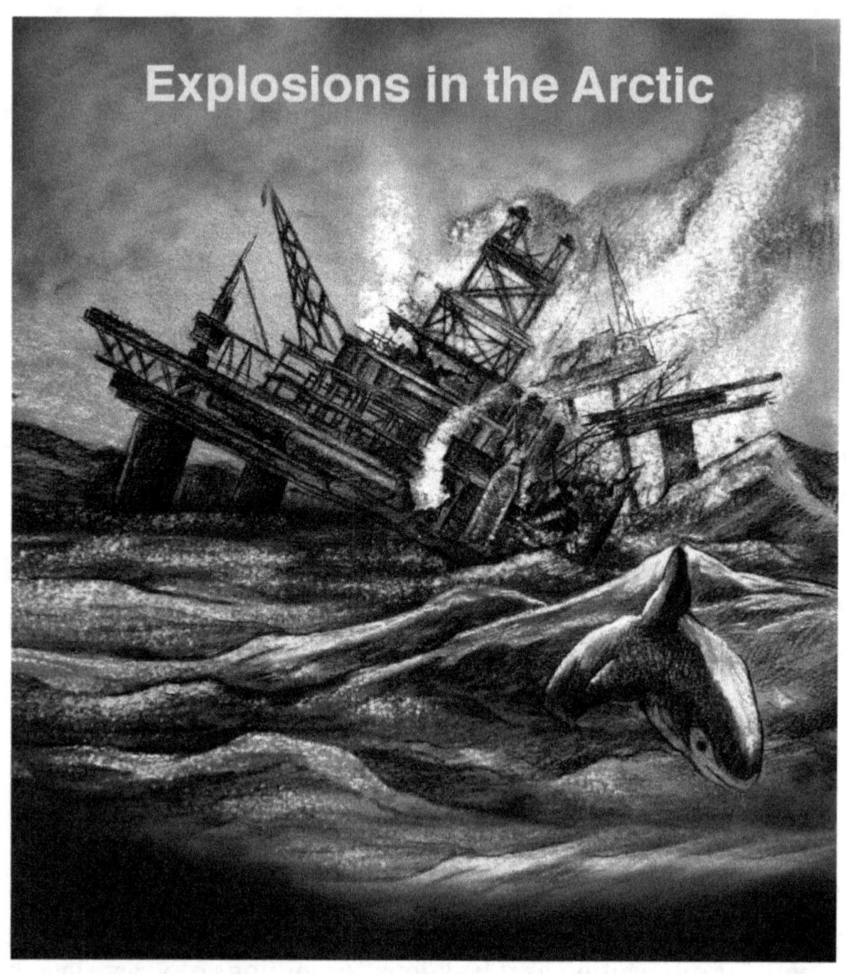

number and decreased the shipment number on the way bill so that it would not be missed when the train arrived at its destination at the Gulf Coast Naval base. It had been stolen and illegally transferred from the train to a waiting truck at mdnight in central Oklahoma. Shipped carefully from the prairie railway crossing through the Middle East (where an arms dealer tripled the price) it was sent on to Russia. Although scheduled for launch six months earlier when construction of the American oil derrick started, its mission had been postponed when transfer to the northern launching facility was delayed by melting permafrost.

Approaching its destination with uncanny accuracy, it gave only seconds of warning to the 28 targets below, who only had time to blink at an onrushing light that exploded with five megatons of force, ending their lives in a fireball, creating numerous widows and orphans, and setting off secondary explosions. Intense heat, crumpled steel, and smoke enveloped the structure, built by engineers to drill, not to withstand the largest explosion humanity could create with such a missile. Mercifully, some died instantly; others when the icy water invaded their bedrooms as the structure quickly sank into oblivion, joining the oil below.

The mood at Exego was sour when the news arrived. "That was my son on that rig," mumbled the company vice president.

In America, where fear mingled with patriotism, oil prices spiked again, the stock market plunged, and as demands for revenge prevailed, counter strikes were launched at Soviet drilling rigs.

Silently, Stormy remembered Jack and wondered how many more lives would be lost in pursuit of oil, or was it really just all about the money?

TERRA, deprived of the carbon emissions that would have come from Arctic oil, simply waited.

Icons to a God

They started appearing in the fields, fifty miles south of Refugia, the new Canadian city established for wealthy climate refugees from the south. Cross-like, made from scraps of burnt wood, ribbons, and faded plastics, they stood like monuments. Each had a hand-like object lifted skyward, and a pool below to catch water. Smoke from burning forests filled the air as pagan-like dancers whirled around the makeshift icons to a god that they did not understand.

"What are those?" asked Stormy.

"They are the prayers of at desperate people," her driver replied. "Those who hope are asking their god to stop the storms, the heat, the lightning, and the strife. We'll see more of them outside of Ottawa, but first we have to stop and recharge the battery. It will only take about 20 minutes."

"How many miles have we gone since the last charge?" Stormy asked.

"About 400," he replied. "These new batteries are great; you don't need gasoline at all."

"It's too bad we didn't all have them 60 years ago," said Stormy. "But the oil companies prevented that."

As she spoke, a very old sport utility vehicle blasted past them at high speed. "I can't believe it!" she exclaimed. "That's a Hummer, one of the gas guzzlers from the early part of the century. Look how big it is, and look at the size of that tailpipe! Oh my God, it's towing a trailer filled with gas cans."

"Most of the gas stations are gone," replied her driver. "He'll need that gas to get to wherever he's going."

Fifty miles later, traffic was interrupted by a violent flash of light, followed by a thunderous explosion. Approaching the scene, Stormy saw the wrecked Hummer and scorched trailer. It had been attacked, blown up by the Climate Posse who were now targeting gas guzzling vehicles on roads across the United States and Canada. *One less vehicle destroying the climate* read their hastily erected sign by the wreckage. Glancing back at her in the rear view mirror, the driver spoke to Stormy. "I thought former American presidents always traveled with a motorcade, body guards, flags, and all that. Why are you traveling alone?"

"Most of the people in the world hate the United States," she replied. "And you know what? I can't blame them. For over a century we kept pumping greenhouse gases into the air, oblivious to the damage it was causing. Tens of millions of people starved while we just kept on doing what we did a hundred years ago; burning coal, subsidizing oil companies, and sending troops to fight for more of it. I tried to change things, but it was too late. By the time I became president there was so much carbon in the air that we were past the tipping point; and then methane kicked in. No wonder so many people want to take it out on me."

"It wasn't your fault," he replied. "China burned a lot of coal, too."

"Yes, they did, but history records that the world's wealthiest nation didn't even pass a climate policy until my administration. Those God damned deniers, those God damned deniers," she muttered as her thoughts trailed off and the little car sped past dead forests.

Speeding past the icons, she wondered, *"How did it come to this? Where are we going? And why?"* She knew the answers. Mankind had been too proud, thinking technology could save them, ignoring the basic processes that control climate, food production, and yes, even interactions between people. Unable to unite, unable to work together, they had continued down a course leading to destabilization, chaos, and massive changes in every one of the environments on planet earth.

"Why are you moving to Canada?" asked the driver.

"Well," she replied, "the rain forests are gone, the Midwestern states, southern Europe, and half of China have become deserts. Coastal cities are flooded, Australia and southern California are bone dry, and the oceans are acid. Where else is there to go?"

"My grandfather used to talk about how the wheat fields turned to dust bowls in Kansas during the 1930s," the driver said. "It's the same now, but this time the rains won't come back. And I remember when we had tuna, salmon, cod, crab, oyster, and shrimp fisheries. Now all we have are massive colonies of jellyfish, dead zones, and algal scum."

"When the oceans are dying, can the land be far behind?" she asked.

A Birthday in Refugia

Stormy blew out 75 candles on her cake, smiled, and cut a small slice. Retired in Refugia, she had escaped the planetary ravages of climate change that had claimed her husband, son, and granddaughter. She had worked enough hours for two lifetimes and now she was weary. The new city sounded a bit cramped and regimented, but a welcome relief from the power blackouts, water shortages, and political chaos of Washington. Residents were promised a reliable water supply, and the complex had its own farm as well. Little was left to do, but she continued to follow the decline of civilization from her little rest home room in northeastern Canada.

At dinner one night she realized that it is not the average climate that gets you, it's the extremes, and the extremes had become more extreme. Floods that had occurred once in one hundred year were now striking every few years in many places, with horrible results. Eight to fifteen category four and five hurricanes formed every year, unleashing the devastation on places such as Los Angeles, San Diego, Boston, Tokyo, and Brazil that were formerly immune from such storms. In mid-continent, tornadoes hit monthly during an expanded storm season, ruining remaining crops and killing people with abandon.

But in Refugia, with a climate typical of what formerly existed in central Kentucky, the extreme weather events were tolerable, and she lived in relative comfort. North central Canada did not suffer from the floods, heat waves, and droughts that plagued most of the world. There, supplied by ample fresh water and adequate food, they lived and waited for the end of the world.

Then the phone call came. It was WGMEDIA, calling to tell her that climate refugees were rioting in Washington, DC. Mobs of lawless people were raiding food, energy, and water supplies wherever they could find them. Was Susana safe? Quickly Stormy called her only daughter on the satellite phone.

"Are you all right?"

"Well, you can't buy a box of cereal in the store anymore. They don't have any because the corn and wheat crops failed! But I did use my federal ration coupon to buy a loaf of bread."

"Stock up on that bread," replied her mom.

"You can't. Under the new rationing system we are only allowed to buy one at a time. But I guess we're lucky; much of the world can't even get a loaf of bread."

Stormy frowned and said, "I remember mom and dad telling me about how sugar, butter, and meat were rationed during World War II. There were price controls too, but I never thought it would happen again during your life time. What about the mobs?"

"Don't worry mom: They won't find me here. Even the solar panels are hidden from the road, and my well keeps pumping out good, cool water. The chickens are laying and beans are ready to be picked. I'm independent and can hang out here until those monsters move on. It's just me, the farm, and baby James now, and we're fine."

"You go girl," said Stormy. "Stay low, avoid the masses, and minimize what you need. You'll be okay. Come up to Canada when it's all over."

"I will; love you mom, bye."

And Now the News: Dateline 2076
Although Stormy no longer roamed the world in search of climate news, her legacy lived on as Susana continued her work. Trained by her mother and possessing her instinctive ability to distill news to its real meaning, Susana reported events unthinkable a few decades ago, and shocking in their description of a planet gone mad. In 2076 she wrote:

MILLIONS FLEE CAIRO AND LIMA: Millions of people in Cairo continue to flee the City as flows in the Nile river reach unprecedented lows and groundwater supplies are exhausted. Emergency drinking water is reserved for hospitals and the military, disrupting industry, farming, and sanitation. The situation is similar in Lima, Peru where glaciers that formerly supplied water have all but disappeared.

GOVERNMENTS COLLAPSE: Unable to pay bills or salaries, many local governments in Africa, Indonesia, and the Philippines disbanded this month as their economies crashed. Victims of unrestrained population growth, disease, famine, and drought, they have emptied their treasuries trying to provide food to increasingly restive populations. Grain prices have increased to as much as 50 times normal, but in many areas food is not available at any price.

SEA LEVEL REMEDIES INADEQUATE: Sea walls, levees, and higher port facilities built to combat sea level rise have cost billions, yet flooding in unprotected areas continues unabated and ground water supplies have been contaminated by sea water. Costly dikes built during the last three decades are now inadequate to repel storms and high tides, but money to keep continually raising them higher and higher cannot be found.

OIL WAR CONTINUES: For both sides, the war over oil in the Arctic has gone badly. Unable to agree on ownership of the massive resources, and unwilling to risk nuclear retaliation by attacking mainland areas, the forty year war continues to drain treasuries, fuel inflation, and destroy economies. The net result however, is that the oil remains underground, not in the atmosphere where it would fuel further climate change.

MOB RULE PREVAILS IN HOUSTON, NEW ORLEANS, AND DETROIT: Unable to pay local police and even the military, chaos has ensued, followed by a breakdown in social order. In the end, mob rule prevailed as thousands of refugees ransacked government offices, food stores, and wealthy enclaves. Emboldened, the Climate Posse began attacking natural gas plants and automobile manufacturing facilities. Climate destabilization is causing societal destabilization.

POWER BLACKOUTS ESCALATE: In the United States, power blackouts have become normal as electricity demand surges during heat waves that envelop southern states and hydropower fails in the northwest. Roving bands of homeless people are tapping into the grid, further disrupting electrical supplies.

REFUGEES FLEE CALIFORNIA: They're migrating east as the Central Valley becomes the new dust bowl, wrote Susana. In a reverse of the migration spawned by the great 1930s Midwestern drought, tens of thousands of farmers and farm workers have fled California for wetter climates in the east. Ignored in Nevada, despised in Utah, and shot at in Colorado, they have continued their march across the plains, past withered crops to the Mississippi river. There, in Ohio, Illinois, and Indiana, they seek work on farms struggling to cope with unpredictable rainfall patterns, tornadoes, crippling hail storms, and increasingly common floods.

Lake Mead Dries Up

Susana's reports continued as she listened to a conversation between several water scientists in southern Nevada.

"I've never seen it this low," said the engineer. Standing by the empty basin that had been Lake Mead, his eye scanned the desolate and barren landscape. Five miles away, dust devils swirled across the dry land where once millions of gallons of water had been stored. Buzzards flew overhead.

"There isn't any water to release downstream," he continued. "I don't know what those people in Las Vegas, Phoenix, and Los Angeles are going to do."

"This is what happens when rainfall is below average for sixty years," replied his partner.

"Yeah, this long-term western drought is continuing in the Colorado river basin," said Susana. "Now look, the reservoir is just about empty."

"I heard there is water in the northern Rockies," said the scientist.

"Yeah, and Denver claims it," replied Susana. "I heard they sent the city police force to protect diversion structures."

"I heard that they tried to divert the whole Colorado river," said his co-worker," and that Phoenix and Las Vegas militias have joined forces to fight the Los Angeles water pirates."

"They used to just argue about it in court but now they're shooting at each other."

"Legal briefs, gunfire, bond measures that fail; none of it will make more rain fall. This is really desperate."

"Well, at least we've got beer; so drink up. Why doesn't Washington do something about it?"

"What can they do? There is not enough water to go around. Besides, all they seem to care about is the oil war in the Arctic."

"I want to see this for myself," announced Susana to the startled men.

"You mean go to Las Vegas?" replied one of them.

"Sure, let's go."

As they entered the empty city, Susana said good bye to the guys, drove her rented car down empty streets littered with tumbleweeds and stopped in front of Caesar's Palace. Looting had been rampant. Broken windows and crumpled doors led to empty, silent, cavernous gambling halls, stripped of value, and filled with hundreds of smashed slot machines.

"I guess there's no market for thousands of slot machines," thought Susana.

Picking through the broken glass on the floor of what once had been *Aladdin's Feast – A King's Restaurant*, Susana imagined she heard the sounds of long gone customers. The voices however, were not ordering food.

"Hands up," came the order from behind an empty bar. Apparently the casino was occupied by squatters, with guns.

Coolly, Susana slowly raised one hand, but the other slipped into her handbag and found the little, pearl handled 38 caliber pistol.

"Give me all your money," demanded a grizzled old man.

"Where's your gun?" asked Susana.

"Uh, right here," he responded, pointing to his hand hidden below the bar.

"Then let's see it," said Susana, pointing her snub nosed pistol at him.

"Oh shit. Don't shoot. I give up."

Sensing his fear, Susana lowered her weapon and sat down on a nearby chair. "Don't worry," she said. "I just want to talk to you, and hey, while you're back there, pour me a drink."

Relieved, the would be robber complied, quickly setting up two drinks.

"You've done this before, haven't you," said Susana.

"Yeah, I used to work here, before it closed. They didn't even give us any warning; just locked us out, and they still owe me two weeks pay."

"Sorry about that. Here's twenty bucks for the drinks. I'm Susana: I write for WGMedia. How about an interview?"

"Sure. What do you want to know?"

"How did it come to this? I mean, what happened to make Las Vegas a ghost town?"

"It was the water, and the blackouts," he replied. "When Lake Mead dried up, there was nothing they could do. Nobody wants to put money in a slot machine that won't work or gamble in the dark. They tried running generators for a while, but when the tourists left, it all fell apart. Sewage was backing up and the only water available was bottled. Unemployment was 60%. When casino investors realized they were losing money, they bailed out. It happened really fast. One year it was fine, the next it was bankruptcy."

"Wow. I'm sorry that happened to you. One of my grandfather's kids used to work here. I guess he's gone now. But you can't stay here. Right?"

"Yeah, I need to move on. But where? Nevada is bankrupt."

"Tell you what. Come to LA with me. Maybe you can get a job there. OK?"

"Wow, I mean, hell yeah, I'll go."

"Good said Susana, putting away her pistol. And do you mind if I quote you in my blog?"

"Sure."

Back in the office, Susana reported on what she had seen.

LAS VEGAS - AN AMERICAN ICON LOST: Billions of dollars built this city, a mecca for desperate people seeking marriage, money, and escape. But it was climate change that destroyed it. Deprived of water by 60 years of below average rainfall and three critically dry years, combined with power outages when lake Mead generators went dry, and grid-wide overloads that created electrical black outs, tourism, the life blood of the city, stopped. Without air conditioning, residents succumbed to the 125-degree heat and moved away. Fountains once depicted in movies are dry, filled with cactus and scrub weeds. Pushed by the dry winds of change, dust has settled on once bright slot machines. Deprived of money, Las Vegas, like the mining towns before it, became the impossible; a ghost town. But for TERRA, such things are not impossible.

AMERICA'S SOUTHWEST SHRIVELS: Phoenix, long isolated from the poverty of the third world, has finally fallen victim to a changing climate. Local reservoirs are dry in most years and as the railway tracks buckled in the heat, shutting off most of the water supply that had been arriving by train, trucks became unable to supply enough drinking water. Sanitation is a thing of the past and people have reverted to outhouses, further polluting the minimal ground water supply. Slowly at first, then in a rush as property values crashed, the population moved away.

As water became a scarce commodity, survival by rule of force became imperative. Just as in Mexico during the beginning of the century, gangs of lawless thugs have spread their control over the land. The National Guard has been sent to the Arctic, and people are left to defend themselves. Only military rule keeps the situation from becoming hopeless, the Army having been expanded into a nationwide police force, charged with enforcing an increasingly strident list of climate laws and trying to defend a shrinking list of law abiding citizens.

The Mexican border in Texas, southern New Mexico, and Arizona is now a wide-open, lawless zone, controlled by drug gangs, smugglers, and banditos. Ranchers, the last bastion of border security, have abandoned their lands as drought crippled their crops, killed their cattle, and eliminated their ability to survive.

A Decade of Climate Change

Massive changes enveloped civilization as the seventh decade of the twenty-first century drew to a close. Ten percent of the planetary population, mostly in China, India, and Africa, died due to starvation, disease, and war. Although it seemed unbelievable to the wealthier nations, these seven hundred million people had fallen victim to chronic shortages of water and food, diseases such as cholera, dysentery, dengue fever, and many yet to be named, as well as to roving warlords who pillaged and plundered unheeded.

"Predictable," was the response by scientists that long ago had said climate change would overwhelm civilizations ability to deal with the multiple threats of global warming.

In 2078, the Gulf Stream current stopped flowing, cooling northern Europe by several degrees. Although not predicted by the Intergovernmental Panel on Climate Change to occur for at least another century, the massive Atlantic current from the warm Caribbean to the North Sea slowed, and then abruptly stopped. For centuries it was driven by the sinking of saltier water off the coast of Greenland which had pulled water in from the south. But now, massive amounts of fresh water from the melting ice sheet have overwhelmed the salinity gradient, disrupting the downward flow that had created the current. Deprived of warm Caribbean water, England, and northern Europe resisted world temperature increases, slowing the need to adapt to massive changes taking place on other continents.

Elsewhere, chaos reigned. As the world sank into a planet wide economic depression people became more desperate. Governments, increasingly focused on protecting water supplies, highways, and airports from rising sea levels, were unable to protect low elevation housing and farmlands. When darkness fell, unequal proportions of wealth were equalized, often with the loss of life.

TERRA: Abandon Hope All Ye Who Enter Here

So said the sign above The Climate Cafe in Refugia, Canada on a hot, summer day in the year 2075. Entering a darkened room, Stormy was greeted by huge, electronic screens on the walls that depicted satellite images of the Earth below them. Other screens were connected to websites with charts, tables of numbers, and bright, flashing colors. Sounds of thunder, roaring wind, and rain came from speakers while fake lightning flashed from ceiling strobe lights. The whole scene reminded her of gambling casinos that had been popular in Las Vegas before the city became a ghost town when Lake Mead had gone dry. She looked up at the ten-foot-wide screen in front of her.

"Climate Box Score," it said. Nations were listed on the left, and for each one, new records of temperature, sea level height, and storm strength were shown in purple. Most of the numbers were purple. Other screens showed huge, real time, satellite photos of the earth where current hurricanes, heat waves, or severe storms were shown in red. Two blistering red circles menaced the Gulf Coast and enormous, brown dust storms hovered over India while Indonesia was engulfed in smoke. Dark blue masses indicated current hundred year floods where people were now dying. Several green areas in Canada and Siberia represented moderate weather and almost thirty percent of the terrestrial planet was depicted in yellow where areas of extreme drought gripped the land. Orange images flickered where forest fires were burning and black blobs meant refugee crises. Large, gray zones were areas of famine. Light blue belts along coastlines showed where sea level rise had forced people from their homes. Stormy gasped as she realized the global extent of the catastrophes.

Almost a third of land on the planet had become desert and charts showed that 20 percent of the Earth's population were refugees, mostly between the tropics of Capricorn and Cancer where failed governments had been unable to stop famine driven wars by desperate people.

At the center of the room, whose floor was deliberately littered with straw, sat an enormous, illuminated, spinning globe, its colors continually changing from shades of green to various depths of red. Staring as the Earth's continents shifted from green to red, Stormy realized that the colors represented population levels.

Dates faded in and out as the colors changed. For 160,000 years since apes began walking upright, the Earth was green and human souls numbered less than 250 million. A white light flashed, showing the current population of the United States; 300 million. " My God, there are more people living in the U.S. now than there were on the whole planet until Christ was born she gasped."

Green remained the dominant color as the date shifted to 1000 AD and human numbers remained below 300 million. But when the Declaration of Independence was signed, the number had tripled to a billion and pockets of light green infected several areas. When her father was born in 1950, human numbers had more than doubled to 2.3 billion while red blotches spread in Africa, India, and China.

Green disappeared by the time Stormy was born in 2000 and numbers of her species leaped to 6.5 billion, almost tripling in 50 short years. But now, in 2075, it had declined to 6.3 billion. Famine, war, and disease were taking their toll and every land mass was a deepening red.

Along the walls, other screens depicted yesterday's videos of refugees. "Keep moving!" shouted a military guard, waving his rifle as the lines of hollow people filed by. Patrons watched the screens with blank stares, realizing that this could be their fate as well. Stormy gasped as a small child stumbled and did not get up.

The screen in front of Stormy flickered, and then changed into an image of carbon concentrations in the air around the world. Most of the globe was shown in purple as 535, labeled as beyond the danger point. Enormous amounts of methane had been breaking down into carbon dioxide, whose levels continued to rise despite humanities efforts.

Moving the cursor, Stormy found the carbon level when she was born in 2000. It was only 380, but had been rising by two every year, and more recently by three. It had not been this high for hundreds of thousands of years, and certainly not since human civilization had begun. Images next to the chart showed masses of automobiles, oil fired power plants, and burning forests. The connection was obvious.

"Look!" she cried as the next image filled the screen. "My God, look how much methane levels have increased in the last ten years. It's way beyond extreme."

"The permafrost is melting," said a patron. "There's enough methane up north to fry us so unless someone can figure out a way to re-freeze the tundra, a few billion tons of it are going into the atmosphere."

"You can't re-freeze the arctic," answered a subdued Stormy. Suddenly, she remembered that Dr. Henley once told her that over a period of a few decades, methane was twenty times more powerful than carbon in terms of trapping the sun's heat.

Taken back by the images, she then noticed waitresses dressed as fire, melting ice, flood, and drought were serving drinks fuming with dry ice (frozen carbon dioxide) that bubbled gently through the alcohol.

"Please don't swallow the ice cube," said the waitress as she handed Story a blue Curacao beverage.

"Welcome to the Africa room." said a comely waitress dressed like a refugee. Stylistic, brown rags were draped over her shoulder, falling in tatters at her thighs. Long, unkempt hair partially obscured her face and her shoes were torn apart.

Looking around, Stormy realized that the rooms were shaped like the earth's continents. A burned rain forest was the theme in the South America room while floods were the theme in China, India, and Southeast Asia. Only in northern Europe, Canada, and Siberia were conditions shown as suitable for civilization. "I hope this place goes out of business soon," she said, picking up her purse and leaving. Looking back she could not help but notice an American business logo above the door. *Those crass, commercial businesses will stop at nothing to make a dollar,* she thought to herself.

As she walked out the door, carbon and methane levels had never been higher in hundreds of thousands of years, and they would continue to warm the planet for millennia, shifting weather patterns, ocean currents, and wiping out twentieth century history, including the Climate Cafe.

<u>Stormy Dreams</u> - Desperation
"I've got to take action, do something," she dreamed. "It cannot go on like this, but how do we stop it?" She twisted and rolled in her bed until the sheets became a tangled mess with pillows on the floor. Drifting in and out of consciousness, she wondered, did she leave the ceiling light on, or was that an overpowering sun above her? The darkened sky had holes in it where millions of stars had disappeared. Nothing was green; all was brown, black, and gray. The seas fought with the land and everything moved, as if pushed by a giant wind. She tried to steady herself, but her slumber ended when she fell on the floor, thankful to be rescued from the nightmare.

"TERRA, where are you? Help us"

"Your path is set," came the reply. "It is too late."

CHAPTER TEN – DEVASTATION
2080 – 2090

Atmospheric Carbon Level – 560-585 ppm

Average Earth Temperature -65°
Sea Level Rise - +11 feet
Earth's Population - 5.8 billion people

A Different Lifestyle

The planetary revolution was spreading. Having started in the slums of the world, it could now be found on every continent, in every county, and in everyone's city and town. Mob violence ebbed and flowed as civilization struggled to retain a resemblance of civility. But hungry people are seldom civil. Only in cities where martial law had been declared and military rule was superior could people feel safe. Washington, D.C. was attacked as protesters bombed and invaded Congress, cornering the newly elected grandson of Senator Inhuff, and forcing his admission that he was behind the aborted capture of George and a plot to assassinate Stormy. Later, committed to a hospital, he was declared legally insane.

Elsewhere, natural gas wells, auto manufacturing plants, cement plants, and oil refineries were attacked. Unprotected wealthy enclaves were invaded, looted, and burned. It was war, but nobody was winning.

The Oceans were boiling

"There's something out there," said the first mate. His binoculars were trained on a fog bank, directly ahead of the ship. A note of fear penetrated his voice.

"Probably just fog," said the cook, who had come on deck to see the massive cloud in front of them.

"It can't be fog," exclaimed the captain. "Not here; not at this time of year."

"If that's what I think it is, we had better get out of here, NOW," warned Dr. Johannsen, who had chartered the research vessel to investigate recent releases of methane from the sea floor.

"It's getting bigger. It's moving toward us: Full astern," shouted the captain. Slowly the boat stopped, and began to move away.

But it was too late. Fed by a massive release of gaseous methane from the melting ice 600 fathoms below, the cloud engulfed the ship in a choking vapor. On deck the seamen struggled to breathe, gasping for breath. Falling to their knees, each man looked into the eyes of the others, seeing nothing but panic and fear as they began to suffocate. Dr. Johannsen tried to cover his mouth, but it was hopeless. His lungs felt like they were on fire as he desperately sucked in the gas, which was almost devoid of oxygen. Within minutes, dead bodies littered the deck.

"Mayday. Mayday," screamed the pilot into a radio in the wheelhouse as the white, deadly gas covered his view out the window and began seeping past the door.

"What is your position?" crackled the reply. But the only sound heard in response was a violent coughing, followed by a thud when the pilot collapsed.

It took longer below, where minutes passed before the poisonous gas penetrated every part of the ship, including the ship's galley where a burner was left on, warming soup for supper. Methane, in addition to being a potent greenhouse gas, is also flammable.

The tremendous explosion of a cloud one half mile wide was witnessed by nearby ships and recorded on world seismographs. Half way around the world in Los Angeles, scientists leaped to their feet when the needle jumped. "It's off the scale," shouted a technician. "Jesus, it broke the needle! That was a bigger jolt than a nuclear bomb could create. What was that?"

The research vessel, Dr. Johannsen and the entire crew perished in a conflagration such as the world had never seen.

Below, in the murky depths, the shock wave liberated more extensive methane deposits, which boiled up from their warming tomb to join the Earth's atmosphere.

Norwegian scientists blamed the holocaust on an earthquake induced landslide that had liberated melting methane ice deposits along the continental shelf. Unleashed from the sea floor, the gaseous bubbles rushed up to the surface, expanding as they rose.

"These undersea sediments have melted before," said Dr. Winn. "Massive methane releases have been documented during previous periods of global warming, but methane ice reformed during the last ice age. To put it mildly, the ocean's methane gun is loaded, and it is pointing right at us. Hopefully, this is not a prelude of things to come," he continued. "If the ocean warms enough and just a few percent of methane deposits down there are released it could transform the climate into a hothouse within a few years.

And Now the News: Dateline 2088

THE SEAS ARE DYING. So wrote Susana for the World News Network that had bought out WGMedia. "In Norway, localized reports of methane and hydrogen sulfide bubbling up from the ocean floor were met with alarm as oxygen levels in the North Sea reached zero and millions of dead fish littered the shorelines. At least one research vessel has burned, having been directly above an erupting methane deposit which ignited, killing all on board.

RISING SEAS THREATEN UTILITIES: Global sea level rise around the world has reached 11 feet, submerging sewer plants, highways, airports, and thousands of miles of shoreline. Nuclear plants on the coast north of Los Angeles and in Florida are operating intermittently as hastily built dikes failed in recent hurricanes and sea water flooded downed transmission lines. Details are sketchy except in areas where electricity is still available.

FOUR HURRICANES RAKE NEW ENGLAND: It began in June when hurricane Esther roared ashore in northern New Jersey and continued up through Maine. Storms in July and August were larger, driven by increasingly warm oceans and measuring five and then six on the expanded scale. But it was the category seven monster in September that did the most damage, striking just south of Manhattan and leaving the once proud city in ruins."

Hurricanes have also struck New Orleans, Tampa, Miami and other coastal cities which have been largely abandoned as the combined destruction of flooding, heat waves and sea level rise crippled transportation and flooded power plants and water treatment facilities. Deprived of electricity, water supply facilities have shut down and finding unpolluted water for millions of residents has become impossible.

"We will keep walking until we find food and water," said one climate refugee from southern Florida."

LOCAL GOVERNMENTS COLLAPSE: Uncounted local governments have simply shut down, deprived of revenue because survival has become more important than paying taxes and utility bills. Services such as fire suppression, schools, and police response have been essentially abandoned. Mail delivery is sporadic or non-existent. Black markets flourish in a world economy where frightened people continue to hoard resources.

CITIES EMPTY AS FOOD SUPPLIES ARE DISRUPTED: Shortages of food have driven most people away from overpopulated cities. After several decades, famine has become truly global. Half a billion people have died in southern China and India from famine, unchecked disease and the ever present local wars and riots. "In the end, there was nothing anyone could do," wrote a reporter in Beijing.

Stormy's Pen Pal

Although much of the determination had gone out of her, Stormy still possessed an unquenchable curiosity, a will to help others and a strong desire to put words on paper. So when the nursing home newsletter printed a request to restore the old tradition of pen pals, she eagerly accepted and began writing.

"Hi: Stormy here, in Canada. Yes, I used to be the American president, but don't tell anyone. America and other formerly wealthy nations are not always well liked in many places these days. Who are you and how is your life?"

"Hi Stormy: My name is Liskah and I live high on a mountain slope on a farm. My husband, son, and I have goats, chickens, and one cow. Our garden gives us food in the summer that we store in the cold soil underground in the winter. Sheep give us wool for warmth and the old windmill with a generator and battery lights our adobe house at night. This land used to be too high and cold to live upon, but the rains are more frequent now and winters are not so cold like they used to be. How are you managing to survive the changing climate in Canada?"

"Well, Canada isn't what it used to be. The climate now is more like what is was 700 miles south of here. The real problem is with the rest of the world. We can't just order food and have it shipped any more. So we have to rely on local farms and fish from the lakes. There are hundreds of lakes here and people are growing bass and perch in them."

"What are bass and perch?"

"They are a type of fish; really good to eat."

"I've heard about fish, but I never ate one."

"Just look them up on the internet, or see if there is still a book about them in the library."

"We don't have a library. What is the internet? Do fish live there?"

Stormy suddenly realized that billions of people alive now did not have access to knowledge or information that she had taken for granted. With sparse electricity, perpetual warfare, hardly any public expenditures, and a collapsed industrial - manufacturing system, centuries of human experience were being lost. Sadly she wrote back, mindful that her pen pal was living life as it was two centuries earlier.

"If you had electricity and a computer, that machine could access information sent like radio waves. The system doesn't work all the time now, and only in wealthy areas not demolished by wars. Maybe someday it will be available to you."

"Oh we're happy the way we are," replied Liskah. "If we need to know anything about growing the vegetables or taking care of the cow, the older ones here will know. Besides, we're too tired at night to play with a machine, and they must be very expensive."

Thinking about her message from Liskah, Stormy slowly began to understand how much the world was changing. Twenty-first, and even twentieth-century technology was disappearing. The reasons were many and varied. Although a lack of electricity due to shortages of coal and oil to run power plants had shut down the internet in more than half of the world, it was a growing shortage of technicians, engineers, and programmers that crippled the world-wide web in formerly developed countries. Even the Internet Archive (whose building lacked adequate security) in San Francisco had been raided by vandals. Increasingly, the lives of these people were being determined by the need to survive: To survive the ravages of refugees (who often raided wealthy areas) to survive a shortage of food (especially in the winter), and to survive deadly disasters such as floods and storms which now struck with increasing regularity. For those people, computers were a luxury they could not afford.

Impact of the Arctic War

Each side feared to directly attack their powerful adversary, so the skirmishes continued, creating in effect, a stalemate. When one side would erect a drilling platform, it would be blown up by guided missiles. In the end, the protracted conflict prevented anyone from reaching the massive carbon deposits 5,000 feet below the churning waves.

It was a war that would save mankind; save them from burning the billions of barrels of hydrocarbons that could have led to a Venus effect, plunging the planet into a permanent hotbed of molten rock and swirling atmosphere where carbon levels exceeded 900,000. Instead, guarded by weapons and bombs, the oil remained below the sea.

In the end it was civilization's greed and inability to forge an agreement that kept the war going, which prevented exploitation of arctic oil, and saved the planet.

Stormy Writes - Who is to Blame?

In her journal, Stormy wrote:

> *Those people on the earth who were willing to sacrifice have tried to stop the wave of climate change. They reduced their carbon emissions, voted for those who realized how dangerous TERRA really was, and even taxed themselves to provide solar and wind energy to third world countries. Slowly, over several decades, their combined efforts have succeeded in finally lowering carbon emissions. Sadly, carbon remaining in the air from their ancestors and parents continues to trap heat, and combined with methane released from feedback loops, feeds an increasingly angry TERRA.*

> *Dr. Henley has died. If governments were still able to afford statues, there should be one for him as the prophet of climate change. A man before his time, his last peer reviewed article, summarized mankind's battle with TERRA, in which he recorded the continuing struggle to limit the use of fossil fuels.*

Success was achieved, but too slowly. It was too little too late and TERRA, being unable to compromise or limit herself, continued to deliver intense storms, famine, drought and war. These in turn have led to worldwide unemployment, economic collapse, and civil instability. I am an old man he said, and I am sorry for the sins of my generation, passed on to those who must struggle to survive on an increasingly hostile planet. The blame, he continued, must rest with those who put their selfish desires before the needs of others. Those who bought elections, paid for misleading media and deliberately created doubt in the face of science, consensus and truth, are responsible for the destruction of civilization."

<u>TERRA - Full Fury</u>

In Hawaii, off the coast of Norway and in Oregon, TERRA unleashed enormous tsunamis as undersea frozen methane deposits melted, shifting massive underwater mountains, and creating waves that destroyed everything in their path for miles inland. Invigorated by the sudden boost in methane, TERRA'S ability to hold heat in the atmosphere increased dramatically on a planet wide scale. It had been 55 million years since she had been able to hold so much of the sun's energy, but this time the changes were happening in mere decades instead of centuries.

Stricken by multiple calamities, an average carbon level of 560 and a global temperature of 65 degrees, her subjects now numbered 5.8 billion people, a reduction from their peak of 8.3 billion in 2050.

<u>Stormy Dreams - Failure</u>

As the earth continued its descent into hell, Stormy's dreams became more integrated with its demise. Attending church services, the sanctuary suddenly became blurry. An inverted crucifix appeared before the congregation, frightening them so much that many hid beneath the pews. Outside the church, a waning moon appeared in a greenish sky. Stars were only dimly visible in the murky night and more than half of them were missing. A violent, red ocean stared at her, rising higher and higher. Black silhouettes of structures crumbled into dust while books, artwork, and sculpture, the pinnacles of humanity that were created in a time of gentle climate, lay broken on the ground.

"You have failed," said a voice in her dreams. Stormy opened her mouth, trying to answer, but could find no words. Finally, she realized that it had been an impossible task. There had been too many people, too many automobiles and power plants, too many compromising politicians, too many uncompromising environmentalists, and too little self-restraint. World-wide, too many forests had been lost and too much ice had melted. Giving up, she drifted off into a deep sleep.

"TERRA, I'm sorry."

CHAPTER ELEVEN – A CENTURY OF CLIMATE CHANGE
2090-2100

Atmospheric Carbon Level – 585-615 ppm

Average Earth Temperature - 67°
Sea Level Rise - + 14 feet
Earth's Population - 4 billion people

What Happened?

For those wealthy enough to survive in 2090, it was a different world. In China, military rule had been overthrown and chaos ruled the day. Life in Indonesia, in the Amazon rain forest that had become scrub land and in the southern United States that had become a desert, was a quest for food, water, and defense from those who would take it from you. Most farms and ranches in Texas had been abandoned, victim to a climate that resembled the Sahara desert. In London, New York, Washington, D.C., and Shanghai, a fourteen-foot sea level rise had robbed the cities of their financial districts, sewage plants, and water supply facilities. Millions were forced to move inland. Winter, as it had been known for centuries, no longer existed for most of the planet.

As access to material things and the energy to power them declined, the common elements of light, darkness, warmth, cold, fire, water, food, companionship, and shelter had become more dominant. Increasingly, people began to set their schedules to daily and seasonal cycles.

Seventy years ago, it had been different. Mobility was rampant as individuals swarmed over the planet in airplanes, cars, and recreational boats. Gourmet beverages, fish, and other edibles from far off countries were readily available if you could afford them. For half of the world's population, houses were large, comfortable, and numerous. Indeed, the only limit to what you could consume was wealth. Too many had it, and the cries of those that did not were ignored, as were early warnings of a climate gone mad.

Struggling to understand it all, Stormy reviewed her writings.

We used up the earth, she had written. It was here for our taking, and we took it. Except for the oil which remains under the Arctic (guarded by nuclear tipped missiles) petroleum reserves are gone. Soils have been ruined, ground water exploited and climate change ignored. In the end, the atmosphere became our weak link to the planet as, filled with greenhouse gases, it turned against humanity, turning fertile ground into deserts and the glacial water supplies of China, India, and South America into barren mountains. The wars over food began in Africa, spread to Asia, then enveloped India and even the Soviet Union as starving Chinese people crossed the border into Russian wheat growing lands. The wars over oil began in the Middle East and spread to the Arctic. Now, in the twilight of humanity, our ecological greed has come back to haunt us. It is a time of vultures calling, of faith repealed, and the ascendance of a terrible TERRA.

<u>Memories</u>

Although retired in Canada, Stormy remembered life before climate had changed so many things. One night, as she lay tossing and turning, she remembered the music sessions at the pub with Boothby. But as TERRA impacted the lives of more and more people, fewer musicians came.

"The music is dying, Boothby," she said aloud in her dreams.

"So it is," he would have replied.

A smile slowly grew on Stormy's lips as she recalled how the music throbbed, faded, caught on again and again while outside the storm hurled lightning, winds, and torrential rain against the windows. She remembered how George gave her that loving look she knew so well, and how her little girl that twirled around, hand over her head ballet style, stealing the show. The crowd back then had been born into a benevolent TERRA and did not understand that she could change. It was an expression of the apex of human development, soon to be lost.

Inside the pub, all was well, but when the music died, TERRA remained, ready to strike any and all who would venture out. Over the years, TERRA wore down the inspiration, energy, camaraderie, and love needed for the musical harmony. Tuesday's at the pub had become quiet, witness only to the sounds of thunder, rain, and silence. Finally, like most other shop owners on the empty street, Sean's grandson gave up and hung a "CLOSED FOREVER" sign on the door. TERRA was molding a new civilization.

And Now the News: Dateline 2096

> *A STRANGE TIME: Seacoast harbors are filled with sailboats, not just the small kind, but larger, trade vessels that have switched to wind power as oil-based fuels became unavailable. Bicycles are everywhere and cars are abandoned on streets, serving as homes but never moving. In rural areas, away from the wretched cities, horses are becoming the dominant mode of transportation, their only fuel being grass that grows where water is available. It seemed that the whole continent has shifted back two centuries in time; a time with less technology. Nights are dark after midnight (pitch black if there is no moon) as the remaining power plants conserve fuel for daytime use.*

Reports and social commentary continued to document the worldwide decline of civilization:

> *ONLY FOUR BILLION PEOPLE REMAIN ON PLANET EARTH: According to reports by the United Nations, the Earth's population continues to decline. War, disease, and famine have taken their toll. But not only human lives have been lost; along with them has been the technological expertise needed to save the Earth. The underlying cause of these calamities has been*

*the destabilization of the Earth's climate, which affects
everything people do.*

*THIRD HURRICANE HITS LOS ANGELES: The third
hurricane to hit southern California this summer brought
devastation and death to the remaining inhabitants. Hurricane
Marie, an immense category six storm, added to the stress of a
summer-long heat wave and decades-long drought. Only
scattered pockets of people remain in the once teeming city,
most having left when water supplies failed in 2082.*

"I guess you had better not use the name Marie," said Susana. "People
will think of the storm that destroyed Los Angeles when they meet
you."

Stormy didn't reply. At 99 years of age and dependent on oxygen from a
machine, she had experienced a century of storms, each decade worse
than the last. She had seen global temperatures rise by ten degrees and
personally witnessed the devastation that it had caused. She had seen
semi-tropical vegetation flourishing in Siberia, unimaginable storms, and
widespread disease. Deserts had spread across the globe as rainfall patterns
shifted toward the poles. Coral reefs had disappeared as seas warmed
and became acid. The Amazon rain forest was gone, and with it those
whose lives had depended upon it. Famine, pestilence, war, and
continuing climate change stalked the land as ever increasing numbers of
hungry climate refugees spread chaos in every city and town. She had
seen sea levels rise 14 feet as Greenland and Antarctica continued to
melt more rapidly than ever, the waters beneath great ice sheets having
been eaten away from below by warming seas. Cities that had been home
to countless millions of people had flooded, permanently. Nations had
rallied in efforts to stop it, but it was much too little and much too late.
Now, she was ready to die.

The World at War Awards a Prize

Few noticed when the Nobel Peace Prize was awarded to Stormy in 2097. Distracted by the collapse of world stock markets lin economies starved of cash by expenditures for sea walls, security, strife, and disease, the event was mentioned between news about mob rule in Africa and Indonesia.

"Congratulations mom," said Susana, "you saved a planet."

"Nonsense," replied Stormy. "The world is in shambles. Besides, maybe it was Uncle Pierre and the Climate Posse that saved it."

Susana gave her mother a grim smile and replied "come on mom, you know it was the Climate Emergency Act and George's legislation that stopped the rise in carbon emissions. When America finally passed carbon legislation you were able to persuade the rest of the world to cut back their emissions too. Without that, we would be on track to a runaway hothouse and nothing would survive."

"Well, the scientists tell me you're right. Even though carbon levels will keep rising, they probably won't reach 1,000 so the planet should survive. It will get worse for a few more centuries, but eventually carbon levels should begin to decline."

Stormy found comfort in realizing that her work had been more valuable than she had imagined, but remembered the words of Al Gore before his death. Like the peace prize recipient that had preceded her, she now understood that the forces of TERRA were greater than that which words could convey. Responding to the award, she made a final plea to humanity.

Put aside your differences. This is bigger than all of us, and only a combined world effort can save us. We are all in the same boat, and it is sinking. We are falling, but we can't fly. Climate destabilization trumps the profit motive, cares not about political differences, and spares neither the poor nor the wealthy. Only by continuing a massive, concerted world effort can we slow the warming that has engulfed us. Only by sacrificing personal pleasure for the greater good can we tame the terror that threatens us.

Sadly, she recognized that these were actions that should have taken place 80 years ago.

TERRA - Finally Limited

Global average temperatures reached 67 degrees, ten degrees more than at the start of the century when TERRA finally began to relent. In the end it was not only the seemingly futile efforts of Stormy, George, and Susana to limit burning of fossil fuels, it was the decline of civilization and its machines that eventually caused TERRA to stabilize carbon and methane levels. Automobiles had all but disappeared; coal plants were rusting to dust; the burning of forests had slowed, air conditioning and indoor heating became limited by shortages of ever more costly electricity and commercial aircraft did not fly. Slowly, the increase in carbon levels began to level off.

But a very different world remained for its inhabitants. Agriculture had not kept pace with the change. Water supplies, governments, commerce, and indeed the structure of civilization had deteriorated and disappeared. The implements of society were gone. In their place, TERRA demanded that people comply with a new climate. Those who did not would not survive. Those who did continued to live with far less comfort, security, and life expectancy than had their ancestors, but they did survive.

TERRA's rush toward global holocaust subsided and the Venus syndrome was aborted, but it was to be several millennia before she would once again become the gentle, stable, and supportive climate that had allowed humanity to flourish. The human population would continue to decline, reaching one billion souls in 5200, then growing only slowly for another century.

Stormy Dreams - Resignation

As sleep overtook her, TERRA filled Stormy's mind. In frustration she tried to reason with the force that now gripped the earth, but the only answers were lightning, heat, and unceasing rain. Angrily, she tried to fight, but her adversary was nowhere, and everywhere. She tried to escape, but there was nowhere to go. Like most of the other people on

Earth, she was broken, exhausted, and hungry. Abandoning civilization, she focused on survival. Gaunt faces drifted by. The earth became an unrecognizable morass of dead crops, flooded cities, and burned forests ruled by TERRA, who was intent on fulfilling her own destiny. To TERRA, Stormy and each life she took, did not exist.

She awoke briefly as the commentator droned on.

> *In Canada, methane gushing from beneath melting permafrost has ignited a firestorm that is sweeping across the provinces, spreading for a thousand miles from British Columbia to Ontario. Drifting smoke lingers over Scandinavia and has reached as far as Moscow. The fires also brought down power lines and hundreds of Canadian cities are without electric power.*

Just as they had when she was born, the lights in Stormy's home flickered, came back on, and then went out. Deprived of oxygen from the machine that needed electricity, she fell into a deep sleep. Lightning flickered outside and Stormy, witness to a century of climate change, slowly died. Enveloped by TERRA, her remains would join the infinite carbon cycle.

EPILOGUE
2100 – 5200 and beyond

Atmospheric Carbon Level – 700 in the year 4,000; 420 in 5200

Average Earth Temperature – 69° in the year 4,000; 60° in 5200
Sea Level Rise +14 feet in 2100; +50 feet in 4,000; +75
feet in 5200.
Earth's Population - 2 billion in the year 4,000; 1 billion in 5200

IN MEMORIAM
ARCHIVES OF THE CLIMATE HISTORICAL SOCIETY
Zurich, Switzerland
2150

IN MEMORY OF THOSE WHO DIED AND CITIES LOST IN THE GREAT
CLIMATE CONFLAGRATION OF THE TWENTY FIRST CENTURY, WE THE
LIVING, CONSECRATE THIS MEMORIAL. GOD FORGIVE US, FOR WE
WERE UNABLE TO SAVE OUR CIVILIZATION.

Shishmaref, Alaska	Coastal Destruction	2002
Somalia, Africa	Famine, Civil War	2045
Kabul, Afghanistan	Famine, Civil War	2052
New Orleans	Hurricanes Nancy & Victor	2056
New Delhi, Karachi	Nuclear War	2062
Tuvalu	Sea level rise	2070
Barrow, AK	Arctic Oil Wars	2042- 90
Washington, DC	Hurricanes Bebe, Mariah & Tolbert	2064-72
Sacramento, CA	Earthquake, flooding, sea level rise	2074
Shanghai	0.3 meters below sea level	2075
Bangkok	0.6 meters below sea level	2077
Miami	2 meters below sea level	2078
Manila	2 meters below sea level	2080
Las Vegas	Desertification	2081
Beijing	Famine, Civil War	2085
Phoenix	Desertification	2091
Galveston	1 meter below sea level	2094
Mexico City	Famine	2095
Moscow	Famine, Civil War	2100
Los Angeles	Desertification	2120

A Sign

Joshua saw it first, looking up from high in the Himalayan Mountains where his village had been established after the Great Warming. A complete circle of colors, unlike anything he had ever seen or been told about before, had appeared in the sky. Shimmering and glowing, it was not bright, but luminous and round.

Later, back at his shelter, he was afraid to speak of it; thinking that the village would consider him to be sick, or taken by the spirits. But the elders knew that he had seen it, and they summoned him to their circle.

"Joshua, son of Thom, grandson of Elijah, great grandson of James and ancestor of Susana; what have you seen?"

"It was a vivid circle of colors, glowing brightly in the sky; red, orange, yellow, green, blue, indigo, and violet," he said. Why have I been chosen to see this?" he asked.

The elders had seen it as well. It was, they said, a circular rainbow, and a good omen. Joshua, ancestor of Stormy, was glad that he had seen it.

Selene, their village at 8,000 feet high on Mount Everest, had gained life as conditions below became more difficult, constrained by heat. Here there was rainfall, and although the soils were poor, crops could be grown with the help of dung from animals and their own waste. Technology, dependent upon electricity and fossil fuels, was essentially absent. In its place, people reverted to a simpler existence, one not unlike that which had sustained mankind in earlier centuries. Mass production, mass consumption, and massive release of greenhouse gases from fossil fuels had ended. In their place, an agrarian, simpler civilization had evolved.

Although much had been lost, the human genome, that chemical assemblage honed by centuries of evolution, had survived, but barely. Gone were the enormous cities, those along coastlines having been drowned by an ever-increasing rise in sea level, which reached 75 feet above what previous civilizations had ever known. Gone were the billions of devices that converted fossil carbon to greenhouse gas. Gone were most of the people that had once traveled around the globe in jet aircraft. In its place were small assemblages of survivors, living in rural environments, mostly young, mostly unaware of the lives that their ancestors had led, bequeathing to them a planet damaged by climate change. The species that had dominated the Earth and brought it to its knees had dwindled to a pathetic remembrance of its numbers.

Although greenhouse gases had destroyed the structure of civilization, it had not destroyed the will to live, or their desire to reproduce.

Survival

The population of the earth was reduced to a fraction of its numbers during Stormy's life, but the primates with too big a brain still survived. In isolated, rural areas at high latitudes and elevations, pockets of people found ways to find food, water, and shelter. Most were without electricity; all were without super markets, super highways, gasoline, and automobiles, and other things that are not essential to life. As the year 4000 dawned, humanity had been reduced to two billion people and was hanging by a perilous thread, but people were alive. Without massive amounts of fossil fuels, transportation reverted to sails, horses, and foot travel. Oil, gasoline, and natural gas became scarce, but some investments made in wind, solar, geothermal, and hydropower were still producing electricity.

The Future?

Although the use of coal had stopped by 2100 and runaway warming had been avoided, greenhouse gases that mankind had released from fossil fuels continued to assault the planet for many centuries. Methane levels declined first, over a period of decades. Carbon levels declined over a time span of centuries, but the oceans, covering two thirds of the globe, remained warm, high, and acidified. It would be several millennia before the deep oceans reached equilibrium and then gave up their heat. Global temperatures rose for many centuries, but slowly, imperceptibly (so long the mind of man cannot comprehend) as the deep oceans gradually cooled, the average Earth temperature stabilized and finally began to decline one degree at a time. Finally they reached a point where global climate patterns began to reverse their century's long drift. Imperceptibly, snow began to reclaim the mountain peaks and polar areas, tundra began to freeze, and ice began to once again form around the poles.

Stormy, George, and their children were gone. Yet the idea to which they dedicated or gave their lives lived on. The idea that humanity could limit itself to live within the natural boundaries of a finite planet was a question that was being continually answered.

> *Hey look how far we've come. Do we know who we are?*
> *Stranded on a mountain top, trying to catch a falling star.*
> *Here's to what we left behind us, here's to what we kept inside.*
> *May the road that lies before us, lead to a place where love abides.*
>
> *Tom Russell: The Man From God Knows Where*

THE END

AFTERWORD

Carbon: The Insidious Threat

Carbon dioxide is invisible, odorless, and colorless. We cannot see it, feel it, or know that it is there. Yet this common chemical is the thermostat that controls temperatures on not only Earth, but also Venus, Mars, and probably other planets. Created from the elements carbon and oxygen, it exists in minute quantities in the air we breathe, the air we use for combustion to create energy, and the air used by plants to create our food. It is the most important element of life.

Although only tiny amounts of carbon are in the atmosphere, it is the fourth most common element in our galaxy and truly enormous quantities exist on our planet, mostly locked up in fossil fuels such as oil and coal that took millions of years to create. By burning these fuels, we are releasing billions of tons of carbon dioxide every year, at a rate 20,000 times faster than has ever happened in the history of our planet. As it enters the atmosphere, we are turning up the thermostat, but since carbon remains in the air for about a century, and since there are no cost effective, politically viable ways to remove it, we have no way to turn down the thermostat.

The Development of Civilization

- The rise of humanity: When the industrial revolution began in 1870, there were about 1.5 billion people on Earth. A century later in 1970, there were 3.7 billion people living on the planet. In one generation, when Stormy was born in 2000, that number had increased to six billion. By 2050 it is projected to be 8.3 billion. At her death due to a climate induced power failure in 2099, the number of people on the planet will have decreased to four billion, and that number will continue to drop rapidly. The interactions of these people with TERRA is the subject of this novel.

- The Rise of Carbon: In 1870, atmospheric levels of carbon on Earth were 280 parts per million, a level that had been conducive to the expansion of humanity. But as steam engines, coal fired power plants, and automobiles multiplied, this level began to rise; slowly at first, then in about 1970, much faster. When Stormy was born in the year 2000, these levels had risen to about 380, which prominent scientists said were causing major changes in our climate. The last time carbon levels were this high, sea level was about 80 feet higher than it is now. When Stormy died in the year 2099, carbon levels had risen to 600, a level higher than it has been in the most recent 650,000 years of Earth's history and one that could begin to change the planet into an inferno of heat, much like the planet Venus is now. On Venus, even on the dark side of the planet, it is so hot that lead melts. Quantities of carbon stored in Earth's coal are quite capable of turning our planet into another Venus.

Carbon levels of 200 are associated with an ice age, when glaciers twenty foot thick covered what is now the city of Chicago. Carbon levels of 600 are associated with nothing that human civilization has ever seen, but which would create a climate in which life in Florida would survive north of the Arctic Circle. At 600, TERRA would not be conducive to the long term survival of civilization in most areas south of the Arctic Circle.

Although our current level of 400 seems like a tiny amount, it exists throughout the entire atmosphere of Earth, and thus the total amount of carbon in the air is enormous. To those who say that such a small amount cannot be important, TERRA asks: Have you experienced the Ebola virus?

• A brief history of the earth's climate: TERRA's history can be read in many places. From deep sea sediments to ice sheets many centuries old; from tree rings to coral rings; from glaciers that trapped carbon, pollen, and dust and in layers of rock formed millions of years ago, those who have studied these things their entire lives can discern the temperature, carbon level and forms of life that lived far before the advent of man. This history tells us that without any carbon in the atmosphere, the Earth would be a snowball, devoid of human life. It also tells us that changes in global average temperature closely mirror changes in atmospheric carbon levels. When carbon levels were low, ice sheets expanded from the poles, creating ice ages. When carbon levels were high, these ice sheets receded and warmer temperatures ruled the Earth.

Much of this history is not written in textbooks written only a few decades ago, for it has only recently come to life. But with the development of modern technology it has been possible to sample ice cores that are miles thick and analyze minute quantities of chemicals. Those of us too old to have learned this in our youth must incorporate this new knowledge into our thinking.

Stable levels, such as have existed for the last thousand years, have allowed our species to multiply and subdue the earth. These levels are no longer stable; you, I and our recent ancestors have seen to that.

Carbon levels rose more rapidly after the start of the industrial revolution, but did not cause as much rise in temperature until about 1980 when clean air regulations began to remove sulfur and other pollutants that had been reflecting heat and keeping the Earth cool.

As of 2013, our Earth's average global temperature has increased by about one degree, which (to the average citizen) does not seem like much. But those who have studied climate tell us that a 4-6 degree increase is likely to happen during the lives of our children. If your body temperature rose 4-6 degrees, you would be in serious trouble; and so will planet Earth.

TERRA's history has recently been shown to include the ability to change much more rapidly than most of us can imagine. For example, Greenland's early European settlers were extinguished by changes that occurred in decades, not centuries. These changes in climate begin slowly, and are thus deceptive. But with acceleration from massive injections of carbon to the atmosphere and spurred on by positive feedback loops, they gain speed and like a runaway train going down a mountain and become impossible to stop. Recent knowledge of this history has shown that TERRA is not stable. Indeed, she is capable of change that we do not want to believe. Ignoring this change will be the downfall of modern civilization.

Stormy is speaking to you. Please listen.

Historians have asked how many people the Earth can hold without destroying itself. But perhaps we should be asking what level of carbon dioxide TERRA can tolerate without destroying us and the planet we have known since civilization began.

Stormy understands these things, and she is trying to tell people of the Earth that such knowledge is crucial to their survival.

Final Thoughts - TERRA'S message to you: Planet Earth has limits. People have become a geological force that can exceed these limits. The level of carbon in the air is a limit beyond which the planet's climate becomes unstable and capable of changing agriculture, sea level, water supplies, and the structure of civilization itself.

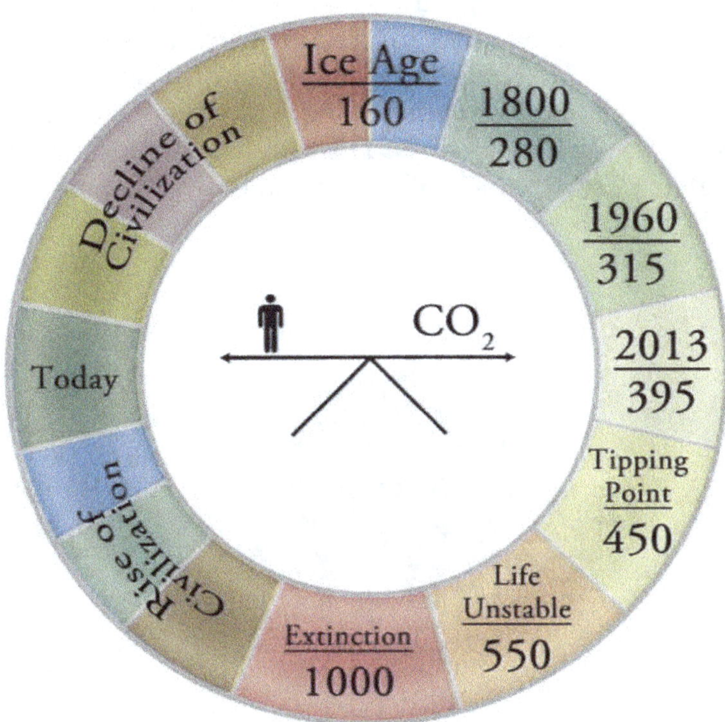

Rotate the see-saw up and down. On the right are average carbon levels in the air. The rise and future decline of civilization for a corresponding carbon level are shown on the left side.

As shown when the CO_2 arrow points right of top center, civilization developed after carbon levels rose above 160 and the last ice age ended. Humanity flourished as levels rose to 280 before the industrial age in 1800 and life continued to thrive as levels rose to 315 in 1960.

Tipping points for runaway climate change are uncertain, but will almost certainly have been exceeded when carbon levels reach 450. When levels reach 550, many forms of life on earth will be unstable. Carbon levels of 1,000 would likely cause the extinction of civilization.

ABOUT THE AUTHOR

Dick Bailey grew up in Chicago, traveled extensively, and studied ecology and natural history, receiving a Doctorate of Forest Resources from the University of Georgia. He has worked in the state and federal government sectors, for consulting firms and in private industry. He currently resides in the San Francisco Bay Area where he directs the Lake Merritt Institute, a small non-profit, community organization and writes about climate change.

www.ingramcontent.com/pod-product-compliance
Lightning Source LLC
Chambersburg PA
CBHW070454120726
47910CB00003B/1040